XXXVI Symposium on Nuclear Physics

(Cocoyoc 2013)

Journal of Physics: Conference Series Volume 492

Cocoyoc, Morelos, Mexico
7-10 January 2013

Editors:

Libertad Barron-Palos
Enrique Martinez-Quiroz
Irving Morales-Agiss

ISBN: 978-1-63266-717-5
ISSN: 1742-6588

Printed from e-media with permission by:

Curran Associates, Inc.
57 Morehouse Lane
Red Hook, NY 12571

CURRAN ASSOCIATES INC. proceedings.com

Some format issues inherent in the e-media version may also appear in this print version.

Printed by Curran Associates, Inc. (2014)

Institute of Physics
Dirac House, Temple Back
Bristol BS1 6BE UK

Phone: 44 1 17 929 7481
Fax: 44 1 17 920 0979

techtracking@iop.org

Additional copies of this publication are available from:

Curran Associates, Inc.
57 Morehouse Lane
Red Hook, NY 12571 USA
Phone: 845-758-0400
Fax: 845-758-2634
Email: curran@proceedings.com
Web: www.proceedings.com

TABLE OF CONTENTS

Journal of Physics: Conference Series **492** (2014) 011001 doi:10.1088/1742-6596/492/1/011001

Preface

XXXVI

Symposium on Nuclear Physics

Cocoyoc, Mexico
January 7-10, 2013

División de
Física Nuclear
SMF
Sociedad Mexicana de Física

The XXXVI Symposium on Nuclear Physics, organized by the Division of Nuclear Physics of the Mexican Physical Society, took place from 7–10 January, 2013. As it is customary, the Symposium was held at the Hotel Hacienda Cocoyoc, in the state of Morelos, Mexico.

This international venue with many years of tradition was attended by outstanding physicists, some of them already regulars to this meeting and others who joined us for the first time; a total of 45 attendees from different countries (Argentina, Brazil, Canada, China, Germany, Italy, Japan, Mexico and the United States). A variety of topics related to nuclear physics (nuclear reactions, radioactive beams, nuclear structure, fundamental neutron physics, sub-nuclear physics and nuclear astrophysics, among others) were presented in 26 invited talks and 10 contributed posters.

Local Organizing Committee

Libertad Barrón-Palos (IF-UNAM)
Enrique Martínez-Quíroz (ININ)
Irving Morales-Agiss (ICN-UNAM)

International Advisory Committee

Osvaldo Civitarese (UNLP, Argentina)
Jerry P Draayer (LSU, USA)
Alfredo Galindo-Uribarri (ORNL, USA)
Paulo Gomes (UFF, Brazil)
Piet Van Isacker (GANIL, France)
James J Kolata (UND, USA)
Reiner Krücken (TRIUMF, Canada)
Jorge López (UTEP, USA)
Stuart Pittel (UD, USA)
W Michael Snow (IU, USA)
Adam Szczepaniak (IU, USA)
Michael Wiescher (UND, USA).

List of Participants

Eli Aguilera (ININ, Mexico)
Paulina Amador-Valenzuela (UAEM, Mexico)
Enrique Barradas-Guevara (BUAP, Mexico)
Libertad Barrón-Palos (IF-UNAM, Mexico)
Adnan Bashir (UMich, Mexico)
Mary Beard (UND, USA)
Tatyana Belyaeva (UAEMex, Mexico)
Roelof Bijker (ICN-UNAM, Mexico)
Francisco Javier Chaparro-Vega (IPN, Mexico)
Efraín R Chávez-Lomelí(IF-UNAM, Mexico)

Journal of Physics: Conference Series **492** (2014) 011001 doi:10.1088/1742-6596/492/1/011001

Osvaldo Civitarese (UNLP, Argentina)
M Scott Dewey (NIST, USA)
Alessia Di Pietro (INFN-LNS, Italy)
Silvia Díaz-Gómez (ICN-UNAM, Mexico)
Alexis Diaz-Torres (ECT, Italy)
J Francisco Favela (IF-UNAM, Mexico)
Olga Félix-Beltrán (BUAP, Mexico)
Ruben Fossion (INGer/C3-UNAM, Mexico)
Alfredo Galindo-Uribarri (ORNL/UT, USA)
Hugo García-Tecocoatzi (ICN-UNAM, Mexico)
Adam Garnsworthy (TRIUMF, Canada)
Paulo Gomes (UFF, Brazil)
Arturo Gómez-Camacho (ININ, Mexico)
Félix Gónzalez-Canales (BUAP, Mexico)
Jorge G Hirsch (ICN-UNAM, Mexico)
Sergio Lerma (UV, Mexico)
David Lizcano (ININ, Mexico)
José Agustín Lozano-Torres (UV, Mexico)
Marissa Maldonado-Velázquez (IF-UNAM, Mexico)
Enrique Martínez-Quiroz (ININ, Mexico)
Joel Mendoza-Temis (GSI, Germany)
Alfonso Mondragón (IF-UNAM, Mexico)
Irving Morales-Agiss (ICN-UNAM, Mexico)
Daniel Julián Nadér (UV, Mexico)
María Esther Ortíz-Salazar (IF-UNAM, Mexico)
Elizabeth Padilla-Rodal (ICN-UNAM, Mexico)
Guy Paic (ICN-UNAM, Mexico)
Jorge Piekarewicz (FSU, USA)
Stuart Pittel (UD, USA)
Víctor Olenin Ramírez-Beltrán (UAEMex, Mexico)
Carlos Alberto Ramírez-Medina (UAEMex, Mexico)
Francisco Ramírez-Sánchez (UAZ, Mexico)
Andrés Ramírez-Morales (IF-UNAM, Mexico)
Tomohiro Uesaka (RIKEN, Japan)
Shan-Gui Zhou (ITP-CAS, China)

Libertad Barrón-Palos
Enrique Martínez-Quíroz
Irving Morales-Agiss
Editors

Journal of Physics: Conference Series **492** (2014) 011002 doi:10.1088/1742-6596/492/1/011002

Peer review statement

All papers published in this volume of *Journal of Physics: Conference Series* have been peer reviewed through processes administered by the proceedings Editors. Reviews were conducted by expert referees to the professional and scientific standards expected of a proceedings journal published by IOP Publishing.

Elastic scattering induced by halo nuclei

A Di Pietro[1], P Figuera[1], M Fisichella[1,2], M Lattuada[1,3] and M Zadro[4]

[1] INFN-Laboratori Nazionali del Sud, via S. Sofia 62, 95125 Catania, Italy.
[2] Centro Siciliano di Fisica Nucleare e Struttura della Materia, Viale A. Doria, 6 Catania Catania 95125 Italy.
[3] Dipartimento di Fisica ed Astronomia, Universitá di Catania, Viale A. Doria, 6 Catania Catania 95125 Italy.
[4] Division of Experimental Physics Ruđer Bošković Institute, Bijenička cesta 54, 10000 Zagreb, Croatia.

E-mail: dipietro@lns.infn.it

Abstract. In this paper an overview of different results concerning elastic scattering in collisions induced by halo nuclei will be given. Due to the very low binding energy of such nuclei, coupling effects with the break-up channel have been found to be very important. Moreover, due to the long tail of the density distribution of halo nuclei and the presence of a large low-lying dipole strength distribution, long range absorption affects the elastic scattering angular distribution. In the paper the main effects responsible for the observed features in the scattering cross-section will be discussed.

1. Introduction

Nuclear halo is a threshold phenomenon arising from the very weak binding of the last one or two valence nucleons which decouples from an inert core that contains all the other nucleons [1]. The core is compact and thus, due to the combination between the short range nuclear force and the weak binding, the valence nucleon(s) can tunnel out into a volume which goes well beyond the nuclear core into the classically-forbidden region. For these reasons, the single-particle wave function of the valence nucleons in halo nuclei has a long tail, which extends mostly outside the potential well.

Since the first observations, a lot of experiments have been performed in order to investigate this peculiar structure also in the attempt to understand how this structure is affecting the reaction process in collisions induced by halo nuclei (see e.g. [2, 3, 4, 5]). Since elastic scattering is a peripheral process, it can be used to probe the tail of the wave function thus allowing to learn about the surface properties of these nuclei.

Elastic scattering was generally considered a simple process where the internal structure of the colliding nuclei has no or little effect on the outcome of the process. Angular distributions are characterized by two regions, divided by the "grazing angle" or "quarter point angle" which separate the illuminated or classically allowed region from the shadow or classically forbidden region. However, the striking shape of the angular distribution asks for attempts to interpret the data in a more complex way in which the explicit inclusion of coupling effects between relative motion and internal degrees of freedom is required.

Due to the low binding energy of halo nuclei, the coupling not only to bound and unbound

excited states but also to the continuum (break-up) is needed (see e.g. [6, 7, 8, 9]).
Elastic scattering experiments involving halo nuclei have been performed at several energies and on targets covering a wide range of masses. These studies have mainly been performed with the 2n-halo ^{6}He nucleus but, more recently, data have appeared in the literature concerning elastic scattering of halo nuclei different than ^{6}He. Many of the measured elastic scattering angular distributions differ deeply from the one induced by the corresponding non-halo isotope. A common feature observed in the low-energy halo-induced elastic scattering, on medium mass and heavy targets, is a reduction of the Coulomb-nuclear interference peak (e.g. [4, 10]) often called in the literature Coulomb "rainbow". The suppression of the Coulomb-nuclear interference seems to be attributed to long range absorption which smooths the surface-jump of the scattering matrix $|S(l)|$. A further consequence of this suppression is a much larger total-reaction (TR) cross-section with respect to the one measured in reactions induced by the corresponding non-halo isotope (e.g. [10, 11]).
From a theoretical point of view, most of the experimental data available in the literature have been analyzed either within the Optical Model (OM) or within the Coupled Channel (CC) formalism. In the OM framework, in order to include the effects of coupling, a Dynamic Polarization Potential (DPP) has been added to the bare potential; the latter has been obtained either phenomenologically or via a double folding procedure. As mentioned above, in collisions induced by halo nuclei, coupling with the continuum is important. Since the coupling with the continuum originates a repulsive DPP (see e.g.[12, 13]), the usual threshold anomaly in the optical potential (OP) [14], which arises from the presence of an attractive DPP, may disappear. A new type of dependence of the OP with the energy, characterized by the absence of the usual threshold anomaly, called in the literature "*breakup threshold anomaly*" (BTA) shows up (see e.g. [15, 16, 17]). Weakly bound halo systems are characterized by a low-lying large dipole strength just above the break-up threshold, it has been found that coupling with this B(E1) strength is important in heavy targets.
In the CC formalism coupling to the continuum is introduced by discretizing the continuum, for each partial wave, from the break-up threshold up to a maximum excitation energy in a finite number of bins evenly spaced in the linear momentum. These type of calculations are called Continuum Discretized Coupled Channel (CDCC) calculations see e.g. [18].
In this paper, an overview of the experimental results present in the literature will be given. In the second section a survey of the elastic scattering results on light targets, obtained both below and well above the barrier, will be presented. In the third section will be presented the results of scattering on medium mass targets and, in section 4, the results that have been obtained on heavy targets.

2. Scattering on light targets

The ^{6}He+^{9}Be elastic scattering has been studied in [19] at E_{lab}=16.8 MeV and in [22] at 16.2 and 21.3 MeV. In [19], an OM analysis was performed using as starting OM parameters the ones from existing ^{6}Li+^{9}Be data and varying the depth of the real and imaginary potential. This analysis did not lead to a good agreement with the data. In [22], in order to investigate the effect of target and projectile excitation, CC calculations were performed where coupling to low-lying rotational states of ^{9}Be and the low-lying 2^+ state in ^{6}He were considered. The agreement with the data improved with respect to the no-coupling case, however a much better agreement with the data, especially at the highest energy, was obtained when coupling to the continuum was included using CDCC calculations. This was also the case for [19]. In order to study the effect of coupling in terms of a DPP, in [22], a trivial local DPP (TLP) was deduced using the prescription of [23]. The extracted real part of the TLP was found to be repulsive at a distance around the sum of the projectile and target radii, whereas the imaginary part has a diffuse tail. These features are similar to the ones found in collisions induced by weakly bound

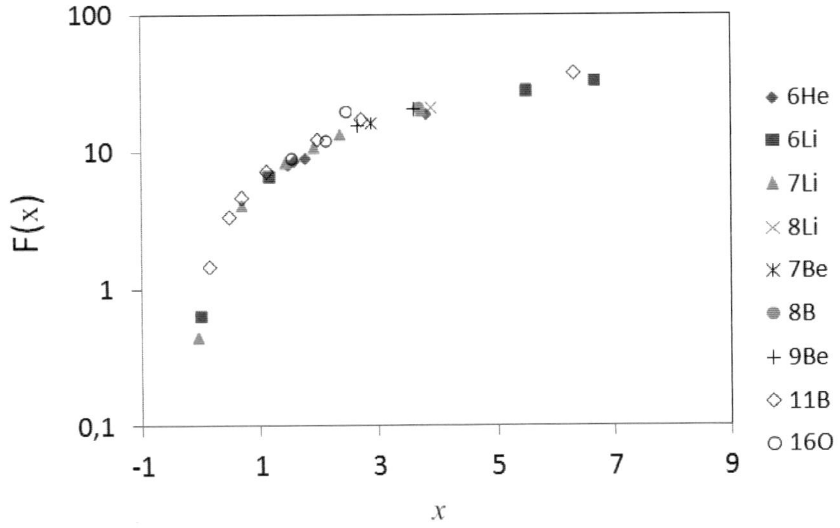

Figure 1. Reduced reaction cross-section for different collisions on ^{12}C targets. Data are from [30] and ref. therein. See text for details.

nuclei and that lead to the occurrence of the BTA, discussed in the introduction.
Contrary to what happens with heavy targets, on light targets the effect of Coulomb break-up is negligible, the effect of coupling on the angular distribution is almost entirely due to nuclear couplings. Another difference that has been observed between light and heavy targets is that, in order to reproduce the data, the inclusion of coupling to the quadrupole 2^+ states in the 2n-α continuum, in the case of ^{6}He, is important while for heavy targets the main effect comes from the coupling with the Coulomb dipole strength [22].

Elastic scattering angular distributions of ^{6}He on ^{6,7}Li and ^{12}C have been reported in [24, 25]. In the analysis of the scattering on ^{6,7}Li targets, global OM potential parameters obtained for the reactions induced by ^{6,7}Li beams were used while, for the scattering on ^{12}C, the potential parameters used were the ones of ^{6}Li+^{12}C at similar energies. Contrary to what found in [19] a good reproduction of the data was obtained in these cases. In order to see possible effects of the halo structure of ^{6}He, the radius of the imaginary potential was increased by 10%. This change produced an overall disagreement with the data. The ^{6}He+^{12}C data were analysed also by Matsumoto et al. [8], using the four-body CDCC approach. A large effect due to the coupling with the four-body break-up is found in these calculations. Moreover, a 15% enhancement of the TR cross-section, with respect to the cross-section of ^{6}Li+^{12}C at a similar energy, is found. In these calculations the Coulomb break-up is neglected.

^{6}He+^{12}C elastic scattering data at much higher energy, 38.3 MeV/nucleon, are reported in [21]. In their analysis OM calculations were performed which required a complex DPP potential, that was added to the double-folding bare potential. The DPP had a repulsive real part in order to account for effects due to the break-up. Four-body CDCC calculations were also performed in [8] which were able to reproduce very well the data once the four body break-up was introduced. Contrary to what found at lower energy , these calculations predict a suppression of the ^{6}He+^{12}C TR cross-section when compared to the ^{6}Li+^{12}C one.

In [32], ^{11}Be+^{12}C scattering data at 49.3 MeV/nucleon were analysed within the OM using as bare potential the one extracted from the ^{10}Be+^{12}C scattering and an analytically derived complex surface DPP which was evaluated from the calculation of the break-up probability.

By assuming an exponential tail for the imaginary surface potential, its diffuseness is directly related to the decay length of the halo wave function. In the case under investigation the obtained diffuseness is 3.2 fm and it depends only on the projectile characteristics and not on the target. A similar diffuseness was in fact obtained in [10] on a different target, as will be discussed in the next section.
In [26] the elastic scattering of ^{6}He on ^{27}Al was measured. The data are rather limited, however the total reaction cross-section was deduced at four energies ranging from 9.5 MeV to 13.4 MeV. The extracted total reaction cross-sections were compared with the ones of stable weakly-bound nuclei (^{6,7}Li and ^{9}Be) and well bound ^{16}O nucleus on the same ^{27}Al target. It was found that, within the errors, the total reaction cross-section of ^{6}He behaves in a similar way as the weakly-bound nuclei and that is larger than for the well bound closed shell ^{16}O.
Elastic scattering of the p-halo ^{8}B nucleus on a ^{12}C target was measured by [30]. CDCC calculations were performed which showed that the effect of break-up coupling for this system is very weak. Moreover in [30], the TR cross-section was deduced and compared with the total reaction cross-section for collisions on the same ^{12}C target induced either by halo (^{6}He,^{8}B), weakly-bound (^{6,7,8}Li, ^{7}Be) and well bound (^{10}Be, ^{11}B, ^{16}O) beams. It was found that, contrary to what concluded in [26], the total reaction cross-section for all these systems follows the same trend. In fig. 1 this comparison is shown. In this case, for matter of comparison, the cross-sections were reduced, in order to eliminate trivial static effects, according to what suggested by [31], where:

$$x = \frac{(E - V_B)}{\hbar\omega}; \tag{1}$$

and

$$F(x) = \frac{2E}{\hbar\omega R^2}\sigma_R; \tag{2}$$

3. Scattering on medium mass targets

As mentioned in the introduction, strong Coulomb coupling effects are expected to affect the elastic scattering angular distribution when the atomic number of the target nucleus (Z_T) is large. As the charge of the target increases, the breakup of the weakly bound projectile is expected to increase as well due to the enhanced importance of the Coulomb break-up over the nuclear one (see e.g. [28]). As mentioned in the previous section, on light targets break-up reactions are dominated by nuclear effects whereas on medium mass targets, both Coulomb and nuclear break-up contribution are important, as well as their interference effect [29]. Similar behavior could also occur for the coupling with the break-up channel.
Very few experiments have been performed of elastic scattering of halo nuclei on medium mass targets. In [11, 27] the elastic scattering ^{6}He+^{64}Zn was measured at 10 and 13.6 MeV and compared with the scattering of the well bound ^{4}He nucleus. From this comparison two main features can be observed: 1) the total reaction cross-section in the ^{6}He case is much larger than in the ^{4}He one ($\approx$ a factor of two); 2) the Coulomb-nuclear interference pattern is reduced. In fig. 2 this comparison at E_{lab}=13.6 MeV is shown.

The large Coulomb dipole break-up probability of ^{6}He and the strong coupling of this process to the elastic scattering could be responsible for the deviation of the ^{6}He scattering from the classical diffraction pattern. This was better investigated in [9] in the study of the scattering induced by a different halo nucleus onto the same ^{64}Zn target: the ^{11}Be nucleus. ^{11}Be is more weakly-bound than ^{6}He and thus has a more extended halo.
In [9, 10], together with the ^{11}Be+^{64}Zn, the ^{10}Be+^{64}Zn elastic scattering was measured as well. As shown in fig. 3, a strong reduction of the elastic cross-section in the Coulomb-nuclear interference region is observed for the ^{11}Be+^{64}Zn collision, much stronger than in the ^{6}He case.

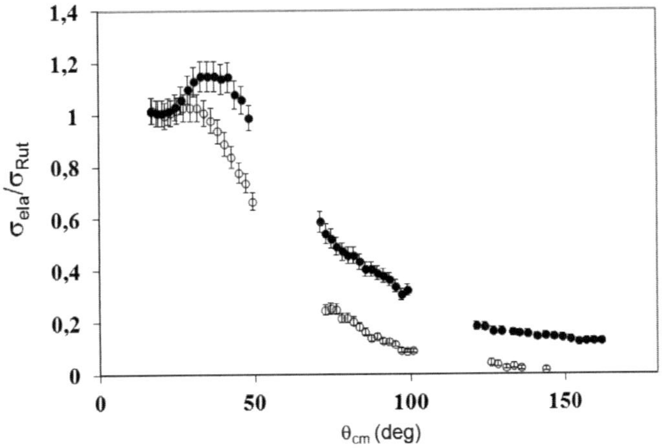

Figure 2. Elastic scattering angular distribution ^{4}He+^{64}Zn (closed symbols) and ^{6}He+^{64}Zn (open symbols) at the same c.m. energy of 12.4 MeV

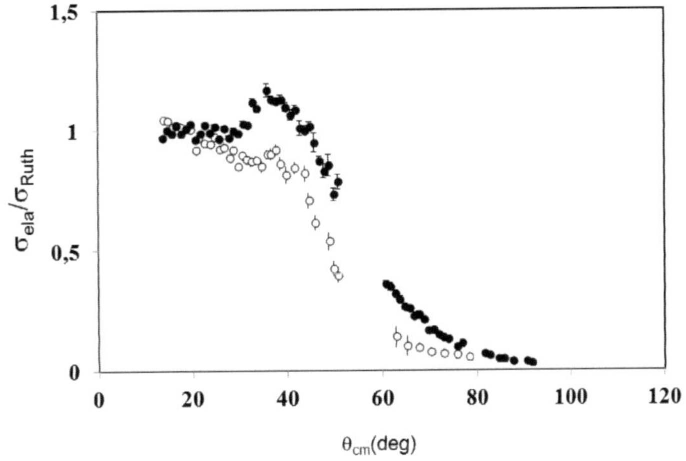

Figure 3. Elastic scattering angular distribution ^{10}Be+^{64}Zn (closed symbols) and ^{11}Be+^{64}Zn (open symbols) at c.m. energy of ≈ 24.5 MeV

In the same figure the ^{10}Be+^{64}Zn elastic scattering angular distribution is shown; this angular distribution presents the typical Fresnel diffraction pattern. The TR cross-section in the ^{11}Be case is ≈ 2.2 times larger than for the ^{10}Be one.
The OP deduced from the ^{10}Be+^{64}Zn measurement was used as bare potential for the ^{11}Be+^{64}Zn calculations, both for the OM and CDCC calculations.

In the OM analysis two types of DPP potentials were used. 1) A phenomenological DPP having the shape of a Woods-Saxon derivative; 2) a DPP potential derived from the semiclassical theory of Alder and Winter for Coulomb excitation which takes into account the effect of the coupling to the inelastic and break-up channels due to the dipole Coulomb interaction. In the first case a good fit of the data was obtained with a DPP having a very large diffuseness of 3.5 fm, in agreement with what found by [32]. In the second case, adding the DPP to the

Journal of Physics: Conference Series **492** (2014) 012001 doi:10.1088/1742-6596/492/1/012001

bare potential produced a strong reduction of the elastic cross-section in the Coulomb-nuclear interference region; however, this reduction is not sufficient to reproduce the measured data. The scattering angular distribution for the ^{11}Be case was also compared with CDCC calculations. In these calculations the ^{11}Be+^{64}Zn reaction is treated within an effective three-body model, ^{10}Be+n+^{64}Zn, and coupling to break-up channels is explicitly taken into account. The calculations show that the suppression of the Coulomb-nuclear interference peak observed in the angular distribution is due to a combined effect of Coulomb and nuclear couplings to break-up channels, although they act in very different way. Coulomb coupling gives a large contribution to the break-up cross section and therefore produces an increase of the reaction cross section; nuclear coupling has a large effect on the elastic phase-shift, thus affecting the shape of the elastic angular distribution but not so much the break-up cross section.
The effect of coupling with the 1n and 2n-transfer processes onto the elastic scattering of ^{6}He on a ^{65}Cu target was investigated in [33] using Coupled Reaction Channel (CRC) calculations. A simplified model for the 2n transfer was used in the CRC. These calculations show that the coupling of the 1n-transfer on the elastic process has a small effect; whereas, the effect of 2n-transfer is much stronger and produces a better reproduction of the data. These calculations however present oscillations which are not observed in the experimental data, the authors attributed this difference to an insufficient absorption.

4. Scattering on heavy targets

In scattering on heavy targets, Coulomb interaction should play an even more important role than with lighter targets. The elastic scattering of the ^{6}He and ^{11}Li halo nuclei from heavy targets shows a strong deviation from the standard diffraction behavior, as the one observed for the scattering of ^{11}Be shown in fig. 3. At low energies, below the Coulomb barrier, the usual Coulomb-nuclear interference peak is completely absent [5, 7, 34, 35]. As mentioned in the previous sections, this behavior is evidence of long range absorption. Also the scattering of the ^{6}Li weakly-bound nucleus shows a similar feature on heavy targets but it is much weaker, being rather a reduction of the Coulomb-nuclear interference than a complete absence, similar to what observed in the case of ^{6}He on the medium mass target ^{64}Zn shown in fig. 2. The elastic scattering of ^{6}He+^{208}Pb has been reported in [5, 34, 35, 36] at energies between 14 and 55 MeV. In [35], and in [34] also in the case of scattering of ^{6}He on ^{197}Au, it was investigated whether at low energies, below and around the Coulomb barrier, the Coulomb dipole polarizability (CDP) is totally or partially responsible for the long range absorption. The CDP is fully dependent on the B(E1) energy distribution and accounts for the coupling with break-up states induced by the dipole Coulomb force acting between projectile and target. The data were analysed within the OM framework. In the OM analysis the CDP was introduced via a DPP. It was found that although the CDP produces an important reduction of the elastic cross-section, CDP can only partially account for the long range absorption found. Beside the CDP, an imaginary potential having large diffuseness ($a_i \approx 1.45$ fm) has to be used. This potential accounts for mechanisms such as higher order effects of the Coulomb coupling or nuclear couplings not included in the CDP. The range of the imaginary potential associated to these additional mechanisms, causing long range absorption, is not as large as the range of the CDP.
In [5], CDCC calculations were performed on the ^{6}He+^{208}Pb at E_{lab}=22 MeV. These calculations confirm the importance of the electric dipole operator in order to reproduce the elastic scattering data.
The ^{11}Li+^{208}Pb elastic scattering has been reported in [4]. Very large reduction of the elastic cross-section has been observed as compared with the scattering of the ^{11}Li core, the ^{9}Li nucleus. Four body CDCC calculations were performed which showed that the elastic cross-section is very sensitive to the dipole strength close to the threshold and in particular to a dipole resonance,

still not well established in the literature, whose energy was adjusted at 320 keV above the threshold, to have the best reproduction of the experimental data. These calculations confirm that also for this system the CDP is the main mechanism responsible for the disappearance of the Coulomb-nuclear interference peak and therefore for the long range absorption.

5. Conclusions

Elastic scattering experiments involving halo nuclei, have been performed at several energies and on different targets over a wide range of masses. In this review a selection of the experimental data available in the literature has been reported. Going through all these data some common features can be pinned down. Many of the measured elastic scattering angular distributions of collisions induced by halo nuclei differ completely from the ones induced by the corresponding non-halo isotope. In the halo nuclei case, coupling with the continuum is found to be very important. For collisions on light targets the coupling is mainly with the nuclear break-up and, as the charge of the target increases, coupling with the Coulomb break-up becomes dominant. An effect of the continuum coupling is the feature observed in the low-energy elastic-scattering on medium mass and heavy targets, i.e. the strong reduction, and in some cases the disappearance, of the Coulomb-nuclear interference peak. This effect is the signature for long-range absorption occurring. A consequence of this suppression is a much larger TR cross-section with respect to the one measured in reactions induced by non-halo nuclei.

In the OM analysis, coupling with the break-up is introduced via a complex DPP potential which has a repulsive real part and a surface imaginary part. By assuming an exponential tail for the imaginary surface potential, its diffuseness is directly related to the decay length of the halo wave function. Therefore, from the scattering measurements, direct information on the wave function of the halo nuclei can be deduced.

References

[1] A. Jensen, K. Riisager and D. V. Fedorov 2004 *Rev. Mod. Phys.* **76** 215
[2] R. Raabe et al. 2004 *Nature* **431** 823
[3] V. Scuderi et al. 2011 *Phys.Rev.* C **84** 064604
[4] M. Cubero et al. 2012 *Phys. Rev. Lett.* **109** 262701
[5] L. Acosta et al. 2011 *Phys.Rev.* C **84** 044604
[6] Y. Kucuk, I. Boztosun and N. Keeley 2009 *Phys. Rev.* C **79** 067601
[7] A. M. Moro, K. Rusek, J. M. Arias, J. Gómez-Camacho and M. Rodríguez-Gallardo 2007 *Phys. Rev.* C **75** 064607
[8] T. Matsumoto, T. Egami, K. Ogata, Y. Iseri, M. Kamimura and M. Yahiro 2006 *Phys. Rev.* C **73** 051602(R)
[9] A. Di Pietro et al. 2012 *Phys. Rev.* C **85** 054607
[10] A. Di Pietro et al. 2010 *Phys. Rev. Lett.* **105** 022701
[11] A. Di Pietro et al. 2004 *Phys. Rev.* C **69** 044613
[12] Y. Sakuragi, 1987 *Phys. Rev.* C **35** 2161
[13] N. Keeley et al. 1994 *Nucl. Phys.* A **571** 326
[14] G. R. Stachler 1999 *Phys. Rep.* **199** 147
[15] A. M. M. Maciel et al. 1999 *Phys. Rev.* C **59** 2103
[16] N. N. Deshmukh et al. 2011 *Phys. Rev.* C **83** 024607
[17] M. Zadro et al. 2009 *Phys. Rev.* C **80** 064610
[18] R. C. Johnson and P. J. R. Soper 1970 *Phys. Rev.* C **1** 976
[19] M. Majer et al. 2010 *Eur. Phys. J.* A **43** 153
[20] T. Matsumoto, E. Hiyama, K. Ogata, Y. Iseri, M. Kamimura, S. Chiba, and M. Yahiro 2004 *Phys. Rev.* C **70** 061601(R)
[21] V. Lapoux et al. 2002 *Phys. Rev.* C **66** 034608
[22] K. C. C. Pires et al. 2011 *Phys. Rev.* C **83** 064603
[23] I.J. Thompson, M.A. Nagarajan, J.S. Lilley and M.J. Smithson 1989 *Nucl. Phys.* A **505** 84
[24] M. Milin et al. 2004 *Nucl. Phys.* A **746** 183c
[25] M. Milin et al. 2004 *Nucl. Phys.* A **730** 284

Journal of Physics: Conference Series **492** (2014) 012001 doi:10.1088/1742-6596/492/1/012001

[26] E.A. Benjamim et al. 2007 *Phys. Lett.* B **647** 30
[27] A. Di Pietro et al. 2003 Europhys. Lett **64** 309
[28] N. Fukuda et al. 2004 *Phys. Rev.* C **70** 054606
[29] R. Chatterjee 2007 *Phys. Rev.* C **75** 064604
[30] A. Barioni et al. 2011 *Phys. Rev.* C **84** 014603
[31] J.M.B. Shorto, P.R.S. Gomes, J. Lubian, L.F. Canto, S. Mukherjee, L.C. Chamon 2009 *Phys. Lett.* B **678** 77
[32] A. Bonaccorso and F. Carstoiu 2002 *Nucl. Phys.* A **706** 322
[33] A. Chatterjee et al. 2008 *Phys. Rev. Lett.* **101** 032701
[34] O.R. Kakuee et al. 2006 *Nucl. Phys.* A **765** 294
[35] A.M. Sánchez-Benítez et al. 2008 *Nucl. Phys.* A **803** 30
[36] N.K. Skobelev et al. 1992 *Z. Phys.* A **341** 315

Journal of Physics: Conference Series **492** (2014) 012002 doi:10.1088/1742-6596/492/1/012002

Near-barrier fusion of proton rich systems

E F Aguilera[1,*], E Martinez-Quiroz[1], P Amador-Valenzuela[1,2], A Gómez-Camacho[1] and J J Kolata[3]

[1] Departamento de Aceleradores, Instituto Nacional de Investigaciones Nucleares, Apartado Postal 18-1027, C. P. 11801, México, D. F., México

[2] Universidad Autónoma del Estado de México, Código Postal 50000, Toluca, México

[3] Physics Department, University of Notre Dame, Notre Dame, In, 46556-5670, USA

E-mail: *eli.aguilera@inin.gob.mx

Abstract. Recent near-barrier measurements for fusion of the proton-halo system ^{8}B + ^{58}Ni (Aguilera E F *et al.* 2011 *Pys. Rev. Lett.* **107** 092701) showed unexpected results. In contrast to previous results for neutron-halo systems, for instance, the respective fusion cross sections are enhanced for both, below- and above-barrier energies. The charged nature of the halo could thus have an effect on the fusion mechanism. Within this context, in the present work the fusion cross sections for several proton rich systems are compared to each other. Three of these systems correspond to the same target, ^{58}Ni (with projectiles of ^{3}He, ^{7}Be, and ^{8}B), but data for ^{17}F + ^{208}Pb also are included. All these projectiles lie near the proton drip line. An appropriate scaling of the data is made to do the comparison and the results are discussed.

1. Introduction

The motivation for the present work is closely attached to the recent results for the fusion of ^{8}B + ^{58}Ni [1]. The main interest of these data comes from the fact that ^{8}B is a proton-halo nucleus in its ground state, and these were actually the first fusion measurements reported for a projectile with such a feature. The data showed a striking enhancement, at all energies, with respect to the expectations for a bare potential. While sub-barrier enhancements are nowdays commonplace, and the mechanisms producing them are rather well established, there is no known mechanism that can produce such a big fusion enhancement above the barrier. In fact, these data fall out of the systematic behavior observed for many weakly-bound systems, including the neutron-halo systems that have been measured. Indeed, the systematics [2–4] shows that, with respect to an appropriate standard reference (see below), complete fusion (CF) of stable weakly bound nuclei is suppressed at energies above the barrier and enhanced below the barrier, while total fusion (TF) for the same systems coincides with said reference at energies above the barrier. As for neutron-halo systems, the measured total fusion cross sections with different targets seem to show a suppression above the barrier and an enhancement at sub-barrier energies.

It is important to consider some issues that might help assess the validity of the observed behavior for the proton-halo system. With respect to the method used to extract the fusion cross sections out from the measured protons, in Ref. [5] arguments are given about the reliability of the procedure. Also, other systems have been studied experimentally with the same method, with quite reasonable results. The fusion data for ^{3}He + ^{58}Ni published in Ref. [6], for instance, were obtained with the same technique. Also, since ^{6}Li was the main contaminant in the ^{8}B beam used in Ref. [1], data for fusion of the ^{6}Li + ^{58}Ni system were obtained simultaneously.

A preliminary analysis of these data indicates quite good consistency with previous results for the similar system ^{6}Li + ^{59}Co [7], which were obtained by another group, using a quite different technique. It is thus important to search for a mechanism that might explain the observed enhancement in the proton-halo system. From the experimental point of view, it is interesting to look at data for other systems that could possibly shed some light on this issue. As a step in this direction, in the present work a comparison is made with similar data for other proton-rich projectiles. The fact that in this context proton-halos seem to behave so differently from neutron-halos, the first showing enhancement while the latter show suppression above the barrier, leads one to think that the charge might be playing an important role.

In next section, the projectiles that will be compared are listed and the method of comparison is described. The results are presented in section 3, where a correlation between the above-barrier cross sections and the breakup threshold energy of the projectile is noticed. A discussion of the results in semiclassical terms is given in section 4 and, finally, a summary and the conclusions of the present work are presented en section 5.

2. Systems considered and method of comparison

Data for five systems were compared to each other, four of them corresponding to proton-rich projectiles and one reference system. Three of the data sets were taken by our own group, using the same technique. These are the already mentioned results for the proton-halo nucleus ^{8}B [1] and newer data for the radioactive nucleus ^{7}Be [8] and for the stable nucleus ^{3}He [6], all of them with the same target, ^{58}Ni. ^{7}Be is actually the core for ^{8}B, which can be considered as a proton weakly bound to this core, with E_{th}=0.14 MeV. Like ^{8}B, the ^{7}Be nucleus lies also on the proton-rich side of the line of stability, and it is weakly-bound, with a threshold separation energy E_{th}=1.59 MeV, corresponding to separation into clusters of ^{4}He and ^{3}He. Being stable, the ^{3}He isotope is an extreme case of proton-rich nuclei; it actually lies on the edge of the proton drip-line, in the light mass end. Also, although maybe one cannot say that it is weakly-bound, its threshold separation energy (E_{th}=5.49 MeV, for separation into $d + t$) is below the typical value for normal nuclei. In the comparison, we included also existing data for fusion of the ^{17}F + ^{208}Pb system [9], whose projectile is actually known to have an excited state with a proton-halo. It is unlikely, however, that the halo state could come into play during the fusion process. Nevertheless, this nucleus certainly has a proton excess and it is weakly-bound too. Its threshold separation energy, into ^{16}O + p, is E_{th}=0.6 MeV. Finally, as a reference, fusion data for the tightly-bound projectile ^{16}O also were included in the comparison, with a ^{58}Ni target [10]. All data sets considered here refer to total fusion, *i.e.*, not only complete fusion but also incomplete fusion, if any, is included in the cross section data.

For the purpose of comparing the fusion data for different systems, we followed the prescription of Canto *et al.* [2, 3]. The barrier parameters $V_B, R_B, \hbar\omega_0$, are obtained from a realistic bare potential and used to reduce the cross section and the energy through the expressions

$$F(x) = \frac{2E}{\hbar\omega_0 {R_B}^2}\sigma, \qquad x = \frac{E - V_B}{\hbar\omega_0}. \tag{1}$$

As a reference to decide whether the data present enhancement or suppression, the reduced cross sections can then be compared with the so called universal fusion function (UFF),

$$F_0(x) = \frac{2E}{\hbar\omega_0 {R_B}^2}\sigma^W = ln[1 + e^{(2\pi x)}], \tag{2}$$

where σ^W stands for the expression derived for the cross section in the one-dimensional barrier penetration model of Wong [11]. With this reduction, it makes sense to compare reduced data for different systems directly in the same plot. To get the barrier parameters, we used always the double-folding São Paulo Potential (SPP) [12], with default values for the matter and charge densities. A summary of the systems to be compared and the respective barrier parameters is presented in Table 1. Note that the data for the ^{7}Be + ^{58}Ni system are still preliminary, but we may say that the points that will be shown are most probably in their final form.

Table 1. Summary of systems analyzed and respective barrier parameters.

System	Reference	R_B (fm)	V_B (MeV)	$\hbar\omega_0$ (MeV)
^{3}He + ^{58}Ni	[6]	8.60	8.59	4.25
^{7}Be + ^{58}Ni	[8]*	8.96	16.58	3.79
^{8}B + ^{58}Ni	[1]	8.92	20.80	4.14
^{16}O + ^{58}Ni	[10]	9.41	31.66	4.44
^{17}F + ^{208}Pb	[9]	11.65	85.32	4.76

* preliminary data

In the sub-barrier region, Wong's formula is inaccurate for systems as light as the ones considered in the present work, which would make the UFF (eq. 2) unsuitable as a reference in this region. To account for this, a renormalization of the cross sections is made [2, 3] by using properly calculated cross sections, σ_{calc}. The first expression of eq.(1) is then replaced by

$$F(x) \rightarrow \frac{\sigma_{exp}}{\sigma_{calc}} F_0(x). \tag{3}$$

To get σ_{calc}, an optical model (OM) calculation was done for each system by using the respective SPP for the real part and an interior imaginary potential of Woods-Saxon form, with parameters W_0 = 50 MeV, r_W = 1.06 fm, a_W = 0.2 fm. The absorption in this potential effectively simulates an incoming wave boundary condition, thus providing a good estimation for fusion.

3. Results

Figure 1 shows the comparison of the reduced fusion cross sections (renormalized) for the five systems of Table 1. The universal fusion function is shown with the dotted line. Thanks to the renormalization performed, this line remains a good reference for the purpose of deciding about enhancement or suppression in the whole energy region.

For the ^{8}B data, the large enhancement already mentioned can be appreciated. An interesting new information is that for ^{7}Be there is also evidence for enhancement above the barrier. This enhancement is not as big as for boron, but it seems to be real. For interpolation purposes, a three-parameter fit using Wong's function [11] was done to each data set, shown with the solid curves. This very reasonable interpolation shows evidence for enhancement above the barrier in the case of the ^{17}F data also, and this enhancement seems to be even higher than that for ^{7}Be. The helium data, on the other hand, seem to show a suppression above the barrier.

A clearer view of the enhancements is obtained by taking the ratios between the interpolation curves in Fig. 1 and the dotted curve, the UFF. The resulting enhancement curves are shown in Fig. 2, which includes only reduced energies above x = -0.5 because this is the region we want to emphasize. For energies far enough above the barrier, these enhancement curves become flat. The ^{16}O curve is a bit enhanced in this region, mainly due to the effect of couplings with inelastic excitations [10]. If one does coupled channel calculations for the boron or the beryllium systems and includes the relevant inelastic excitations, the effects on fusion are negligible. So,

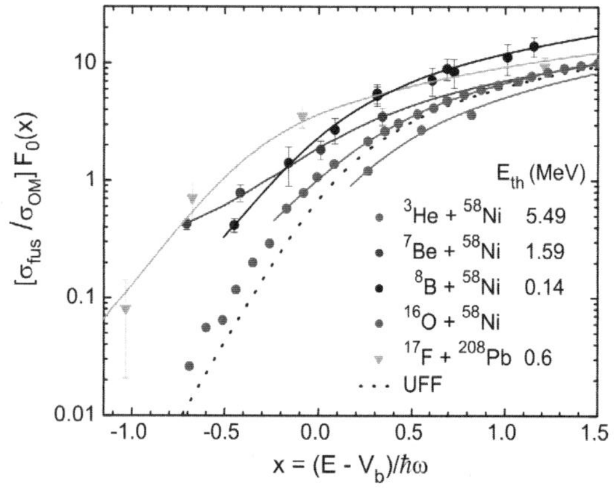

Figure 1. Reduced fusion data for five systems. The solid curves are Wong function fits to the respective data while the dotted curve is the UFF (eq. 2). E_{th} refers to the breakup threshold energy for each projectile.

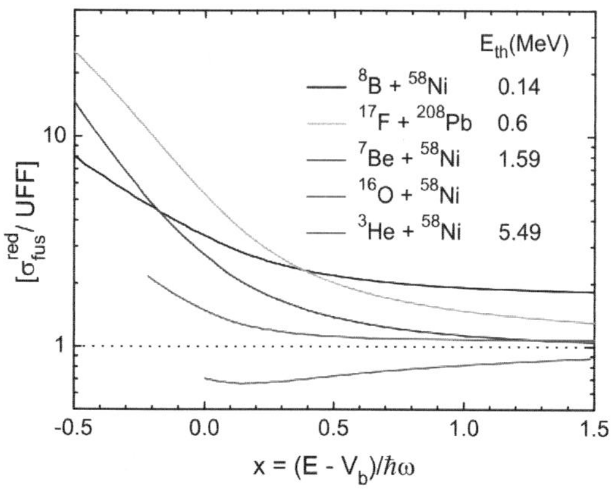

Figure 2. Enhancement curves, with respect to the dash-dotted line of Fig. 1.

a different mechanism is needed to explain the larger enhancements observed for these systems. Indicated in Fig. 2 are the threshold energies (E_{th}) needed to separate each projectile into two clusters. Looking carefully, one may notice a correlation between the observed enhancements, in the region where the curves are flat, and these energies. To further investigate this point, we took the enhancements at x = 1, where the curves are already pretty flat, and plotted the respective values versus E_{th}, as shown in Fig. 3.

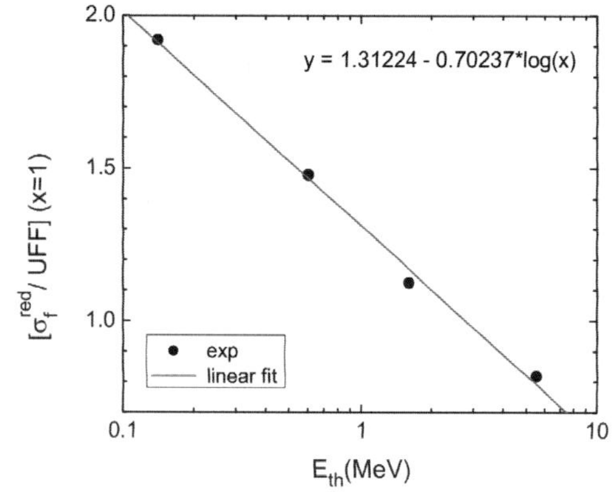

Figure 3. Correlation between the fusion enhancement at x = 1 and the breakup threshold energy of the projectile.

Quite surprisingly, this plot shows a nice correlation between the reduced cross sections at "high energy" and the breakup threshold energies of the projectiles. In the logarithmic scale of the figure, the points lie along a well defined straight line. The question remains as to the physical reasons lying behind this correlation. A complete theoretical explanation of this is beyond the scope of the present work. Instead, some simple semiclassical considerations will be made in next section, which might help to get further insight into this matter.

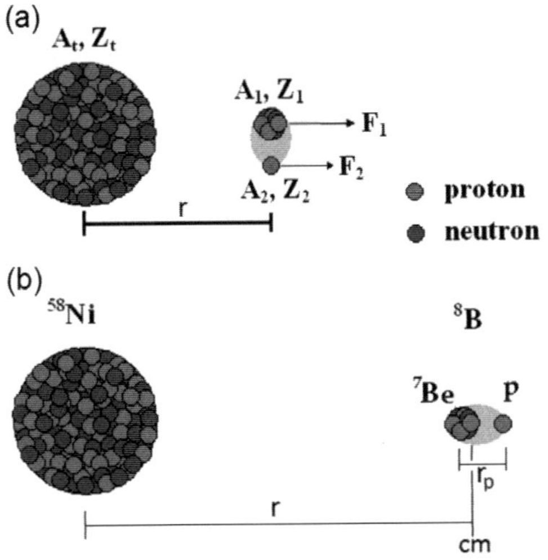

Figure 4. (a) Projectile formed by two nearly uncoupled clusters, exposed to the Coulomb field of the target. (b) Hypothetical fully polarized situation for the case of the ^{8}B + ^{58}Ni system.

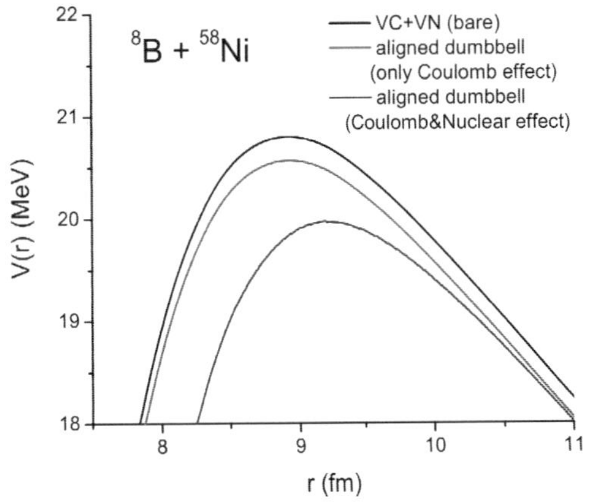

Figure 5. Barrier change induced by the polarization illustrated in Fig. 4b.

4. Discussion

Let's assume temporarily that we have a projectile formed by two nearly uncoupled charged clusters, which approaches some target and gets exposed to the respective Coulomb field (see Fig. 4a). In classical terms, the acceleration on each cluster is easily calculated and its ratio is given in terms of the respective charges and masses, $a_1/a_2 = Z_1A_2/Z_2A_1$. If this ratio is less than one, the light cluster tends to stay behind. Table 2 presents the most probable clusterization of our projectiles, and the respective values of the acceleration ratio. This ratio is less than one in all cases, so according to our classical argument, the light cluster in each case would tend to stay behind. Of course, these classical arguments are suspect for a quantum system, but Esbensen and Bertsch [13] have made breakup calculations for ^{8}B + ^{58}Ni, where the boron is treated quantically, showing that Coulomb polarization is quite important in this process.

While the hypothesis of nearly uncoupled clusters is probably valid for halo nuclei such as ^{8}B [14], it will break down for other projectiles. Intuitively, one can expect that the binding energy of the clusters would play a role in this polarization process: the larger the binding the harder the polarization of the projectile. So, this could be the mechanism underlying the correlation that we observed in Fig. 3. But there is still the question as to how could this mechanism actually contribute to the fusion enhancement.

Suppose that we have our B-Ni system in the fully polarized configuration at a distance r between the respective centers of mass, as illustrated in Fig 4b. We can estimate how the Coulomb barrier is modified in this situation. The interaction potential can be calculated as a superposition of the contribution of the Be cluster plus the contribution of the proton

$$V_{BNi}(r) = V_{BeNi}(r - r_p/8) + V_{pNi}(r + 7r_p/8). \tag{4}$$

For the first part, we used the São Paulo Potential for ^{7}Be + ^{58}Ni. For the p-Ni system, the nuclear potential is negligible in the region near the barrier and for the respective Coulomb part we used the potential of a point plus a spherical charge distribution of the proper radius. As a

Table 2. Projectile clusters and respective separation energies and acceleration ratios.

Proj.	Clusters	E_{th} (MeV)	Z_1	A_1	Z_2	A_2	a_1/a_2
^{8}B	^{7}Be + p	0.14	4	7	1	1	4/7
^{17}F	^{16}O + p	0.6	8	16	1	1	1/2
^{7}Be	^{4}He + ^{3}He	1.59	2	4	2	3	3/4
^{3}He	^{2}H + p	5.49	1	2	1	1	1/2

reasonable estimation, for the radius of the orbit of the valence proton in boron we used $r_p \sim 3$ fm.

Figure 5 shows the obtained results, where the blue curve corresponds to the polarized barrier. Its height is considerably lower than that of the bare (unpolarized) potential, shown in black. The barrier radius is also slightly increased. Both of these effects should produce an enhancement on the fusion cross section. It is interesting to compare the importance of the nuclear and the Coulomb parts in this process of lowering the barrier. If we assume that the nuclear interaction stays the same as in the bare potential and estimate only the change of the Coulomb potential due to the dumbbell configuration, we get the red curve. Clearly, the nuclear interaction plays the main role.

To get an idea about the magnitude of the enhancement induced by the above polarized potential, a barrier penetration model (BPM) calculation using this barrier was performed. The result, shown in Fig 6 (blue curve) is rather encouraging: it almost describes well the data. Further measurements are required to corroborate the observed fusion enhancements and more rigorous theoretical calculations are needed to understand them. Based on the results of this work, it is fair to say that Coulomb polarization might be playing an important role.

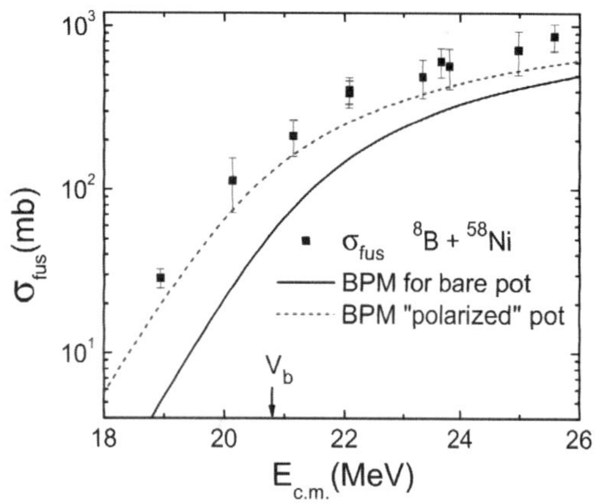

Figure 6. BPM calculations for the bare (solid) and the polarized (dash) potentials of Fig. 5.

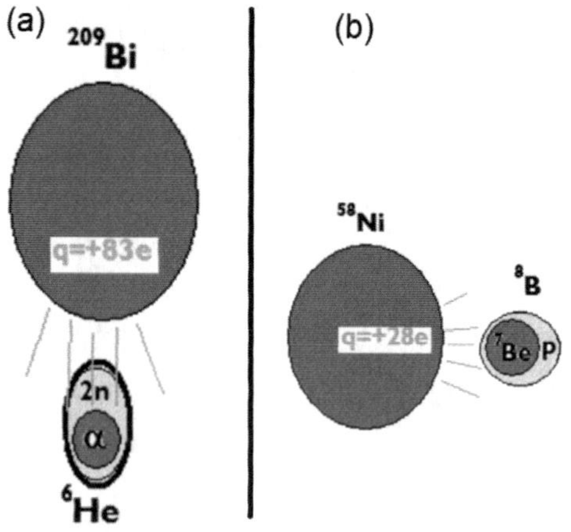

Figure 7. Visualization of the possible role of Coulomb polarization in the reactions involving halo nuclei.

5. Conclusions

The near-barrier fusion cross sections measured for several proton-rich projectiles (^{3}He, ^{7}Be, ^{8}B, ^{17}F) were compared to each other and to those corresponding to a reference projectle (^{16}O). A reduction of data was done to make the comparison, by using the prescription of Canto *et al.* [2, 3] and the São Paulo potential to get the corresponding barrier parameters. Evidence for enhancement above the barrier was pointed out for projectiles of ^{7}Be, ^{8}B and ^{17}F while the opposite holds for the ^{3}He projectile. For energies sufficiently above the barrier ($E \sim V_B + \hbar\omega$), a correlation was found between the reduced cross sections and the breakup threshold energy of the projectile. Simple semiclassical considerations point to Coulomb polarization as the possible mechanism underlying the observed behaviour.

Further pursuing this hypothesis, it is tempting trying to visualize the proton-halo effects observed so far, as opposed to the respective neutron-halo effects, by using the physical picture illustrated in Fig. 7. For neutron-halo nuclei, Coulomb polarization favors neutrons in the halo residing in the region between the core and the target, which then enhances the reaction probabilities. This is consistent, for instance, with the observation that for ^{6}He + ^{209}Bi the main reaction mechanism in the near-barrier region is 2n transfer. In contrast, polarization of the proton-halo projectile via the Coulomb force, results in the valence proton spending more time at large distances from the target, where it is shielded by the core from the full Coulex effect. In this case, core-halo breakup would occur mainly through the long range Coulomb force, and p-transfer would be suppressed. The projectile would "see" a lower barrier, thus enhancing fusion.

Acknowledgments

This work has been partially supported by CONACYT (México) and by the NSF (USA) under Grant No. 09-69456.

References

[1] Aguilera E F *et al.* 2011 *Phys. Rev. Lett.* **107** 092701
[2] Canto L F *et al.* 2009 *J. Phys.* G **36** 015109
[3] Canto L F *et al.* 2009 *Nucl. Phys.* A **821** 51
[4] Gomes P R S, Lubian J and Canto L F 2009 *Phys. Rev.* C **79** 027606
[5] Amador-Valenzuela P, Aguilera E F, Martinez-Quiroz E, Lizcano D, Belyaeva T L and Kolata J J 2013 *J. Phys.: Conf. Series* this Conference
[6] Aguilera E F *et al.* 2013 *Phys. Rev.* C **87** 014613
[7] Beck C *et al.* 2003 *Phys. Rev.* C **67** 054602
[8] Martinez-Quiroz E *et al.* 2013 to be submitted
[9] Rehm K E *et al.* 1998 *Phys. Rev. Lett.* **81** 3341
[10] Keeley N, Lilley J S, Wei J X, Dasgupta M, Hinde D J, Leigh J R, Mein J C, Morton C R, Timmers H and Rowley N 1998 *Nucl. Phys.* A **628** 1
[11] Wong C Y 1973 *Phys. Rev. Lett.* **31** 766
[12] Chamon L C, Carlson B V, Gasques L R, Pereira D, De Conti C, Alvarez M A G, Hussein M S, Cândido Ribeiro M A, Rossi Jr. E S and Silva C P 2002 *Phys. Rev.* C **66** 014610
[13] Esbensen H and Bertsch G F 2002 *Phys. Rev.* C **66** 044609
[14] Aguilera E F, Kolata J J and Acosta L 2010 *Phys. Rev.* C **81** 011604(R)

Journal of Physics: Conference Series **492** (2014) 012003 doi:10.1088/1742-6596/492/1/012003

Mapping σ_p into σ_{fus} for the ^{8}B+^{58}Ni system

P Amador-Valenzuela[1,2], E F Aguilera[1,*], E Martinez-Quiroz[1], D Lizcano[1], T L Belyaeva[2] and J J Kolata[3]

[1] Departamento de Aceleradores, Instituto Nacional de Investigaciones Nucleares, Apartado Postal 18-1027, Código Postal 11801, México, D.F., México
[2] Universidad Autónoma del Estado de México, Código Postal 50000, Toluca, México
[3] Physics Department, University of Notre Dame, Notre Dame, Indiana 46556-5670, USA

E-mail: *eli.aguilera@inin.gob.mx

Abstract. A fusion excitation function for the ^{8}B+^{58}Ni proton-halo system at near Coulomb barrier energies was recently published (Aguilera E F *et al.* 2011 *Phys. Rev. Lett.* **107** 092701). The respective fusion cross sections (σ_{fus}) were determined from the integrated cross sections for the evaporated protons (σ_p), by using proton multiplicities (M_p) calculated with the code PACE. In this code, the transmission coefficients T_l are calculated only for the compound-nucleus values of A and Z and an extrapolation is then made for subsequent decays. In the present work, the reliability of the whole procedure is further investigated both, by studying the effect of variations in the input parameters to PACE and by computing the M_p values with the alternate code LILITA. An explicit calculation of all necessary transmission coefficients is made in the latter code.

1. Introduction

The radioactive nucleus ^{8}B is adjacent to the proton drip line, it is a short-lived nucleus ($T_{1/2} = 0.8$ s) and has a very small proton separation energy of only 0.138 MeV. This radioactive nucleus plays a role in the production of high-energy neutrinos in the sun [1]. In addition, due to the work done in Refs. [2–4], it has attracted much attention in the past decade. The experimental angular and energy distributions of ^{7}Be after ^{8}B breakup on a ^{58}Ni target [2, 3], and the large reaction cross sections observed for the same system [4], provide evidence about the proton halo character of ^{8}B. In this matter, recently the fusion excitation function for the ^{8}B+^{58}Ni system at near Coulomb barrier energies was reported [5], showing an unusual enhancement most probably related also to proton-halo effects.

Much experimental and theoretical effort has been dedicated to investigate nuclear reactions of exotic neutron-halo nuclei [6–8]. The situation is quite different for the proton-halo nuclei since there is not so much information on them. For neutron-halo systems there is a *suppression* of the fusion cross section, σ_{fus}, above the Coulomb barrier. In contrast, the data of Ref. [5] show an *enhancement* of σ_{fus} below and above the barrier, specifically for the ^{8}B+^{58}Ni system.

The respective values of σ_{fus} in Ref. [5] were determined from the integrated cross sections for the evaporated protons (σ_p), by using proton multiplicities (M_p) calculated with the code PACE [9]. In the present work, the reliability of the whole procedure used to determine σ_{fus}, is further investigated both, by studying the effect of variations in the input parameters to PACE [9] and by computing the M_p values with the alternate code LILITA [10, 11].

2. Antecedents

Since the present work relies on the results given in Ref. [5], in this section an insight of the most important details, regarding the experimental arrangement and other considerations related with the determination of σ_{fus}, will be given.

The ^{8}B beams were generated using the ^{3}He(^{6}Li,^{8}B) reaction at the *TwinSol* facility [12] at the University of Notre Dame during four different stages. Primary $^6Li^{3+}$ beams with energies between 31 and 38 MeV produced ^{8}B beams with incident (lab) energies between 22.4 and 30.0 MeV. Primary beam currents for the different stages varied from 50 to 220 particle-nanoamperes (pnA), yielding secondary beam rates of 0.5 - 2.7$\times 10^4$ s^{-1}. In order to detect protons evaporated from the fused system (^{8}B+^{58}Ni), several $\Delta E - E$ silicon surface-barrier telescopes were placed at backward angles that ranged from 112.5° up to 157.5°. At forward angles two additional detectors were used to monitor the beam.

The secondary beam from *TwinSol* is generally contaminated by unwanted ions having the same magnetic rigidity. A bunched beam was used in order to identify the reaction products of interest, and to reject direct protons, by time of flight. Because of their partial time overlap with the ^{8}B beam, the most relevant contaminants and typical yields (relative to ^{8}B) were $^3He^{2+}$ ($\sim$45%) and $^7Li^{3+}$ ($\sim$45%). In order to determine the contributions of the ^{7}Li and ^{3}He contaminant beams to the proton yields, additional experiments were performed to measure protons produced by these two projectiles on a ^{58}Ni target. It turned out that the ^{7}Li component was not a concern because the ^{7}Li beam energies were always well below the Coulomb barrier. This, combined with the fact that proton multiplicities for ^{7}Li+^{58}Ni are about a factor of 2 lower that those for ^{8}B+^{58}Ni, leads to a negligible effect ($\sim$1%) on the proton yields. On the other hand, the involved ^{3}He energies were above the respective Coulomb barrier, with important fusion cross section values [13]. A contribution of up to 10% coming from the ^{3}He beam component was corrected for in a few cases. A typical proton spectrum is shown in Fig. 1. It can be seen that the experimental points closely follow the predictions of the evaporation model, computed with the code `PACE` [9].

As mentioned above, the experiment was performed in four stages. In the first stage a 1.36 mg/cm^2 natural Ni target was used and the bombardment was done at E_{lab} = 27.5 and 30.0 MeV. It is worth mentioning that in the present paper we follow the notation defined in Ref. [5] for the energies. In other words, E_{lab} refers to the lab energy of the projectile before entering the target while E_{fus} indicates the effective energy for fusion in the center of mass reference frame. As mentioned below, E_{fus} includes corrections for energy-loss in the target, properly weighted by the respective fusion cross section. For the second stage, the target was enriched ^{58}Ni (0.924 mg/cm^2) and the measurements were done at E_{lab} = 25.8, 27.6 and 29.1 MeV. In the third stage two targets were used. The first one was an enriched ^{58}Ni (0.725 mg/cm^2) with measurements at E_{lab} = 24.6 and 26.9 MeV. The second target, used for the measurement at E_{lab} = 22.4 MeV, was a natural Ni foil (5.6 mg/cm^2). Finally, a natural Ni target (2.22 mg/cm^2) was also used in the fourth stage. The measurements were done at E_{lab} = 23.9, 26.5 and 28.6 MeV. Background determinations were performed during stages 3 and 4 by using pairs of identical target frames. One of them had no target and was used for blank-target measurements right after the respective runs with a target. No modifications of the setup or the beam conditions were made. In addition to providing the proper corrections for the data taken in stages 3 and 4, these blank-target data showed a smooth energy dependence that was used to estimate a background correction for the measurements of stages 1 and 2.

The evaporation protons were measured at the backward angles, so that it was possible to determine the proton cross section for each angle. Since the respective angular distribution is flat, an integration over the whole solid angle can be performed to obtain the total proton cross section, σ_p. The procedure is illustrated in Fig. 2, where the differential proton cross sections determined for each detector angle are shown for the bombardment at E_{lab} = 30.0 MeV. The solid curve, calculated with PACE, was used to obtain the integrated cross section. Finally, a mapping was done to get the fusion cross section, σ_{fus}, by using the proton multiplicities calculated with the code PACE [9]. The fusion cross sections reported are based on the assumption that complete fusion (CF) is the only mechanism responsible for the observed proton yields. One concern was the possibility of having incomplete fusion (ICF), where the residual ^{7}Be after projectile breakup fuses with the target. For a given energy E, the effect of this process on the extracted cross sections depends on the factor f = CF/(CF+ICF) and on the ratio of the corresponding proton multiplicities $r(E) = M_p(^7\text{Be})/M_p(^8\text{B})$. If $r(E) = 1$ for all E, the reported values of σ_{fus} would correspond to total fusion (TF=CF+ICF). Estimations based on PACE [9] calculations indicate that $r(E)$ is actually an increasing function of E such that, for the experimental energy range, $0.72 \leq r(E) \leq 0.85$. As a result, the effect of ICF is more important at lower energies. A value $f = 0.7$ would have a negligible effect on σ_{fus} for all values of E as long as the fusion cross section is associated with TF. Even a factor as small as $f = 0.5$ should leave the values of σ_{fus} within the reported error bars for all energies except the lowest one, where σ_{fus} would increase by 20% (the respective error bar is 14%). Similar results are obtained for ^{4}He or ^{3}He capture. The correction factors are somewhat larger in these cases, but it is worth noting that the already large cross sections will always be increased if there is substantial ICF since the proton multiplicities for ICF are always smaller than those for CF. Note also that the measured total reaction cross section places an upper bound on σ_{fus}. Therefore, while we cannot rule out the ICF possibility, this will not substantially change the reported values of σ_{fus} as long as they are taken to refer to total fusion.

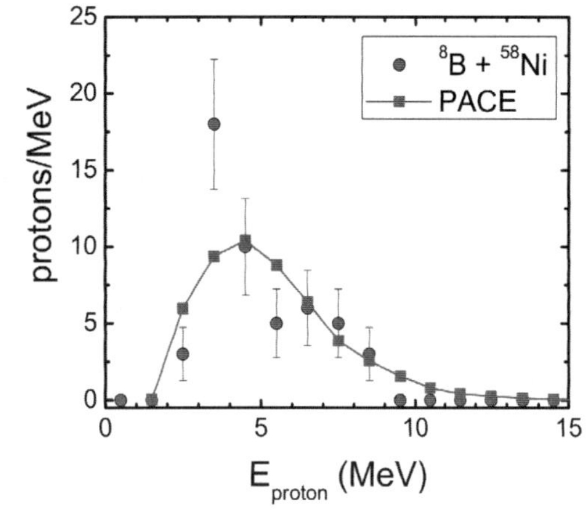

Figure 1. Typical proton spectrum (E_{lab} = 22.4 MeV) compared with a PACE [9] calculation. Data taken from Ref. [5].

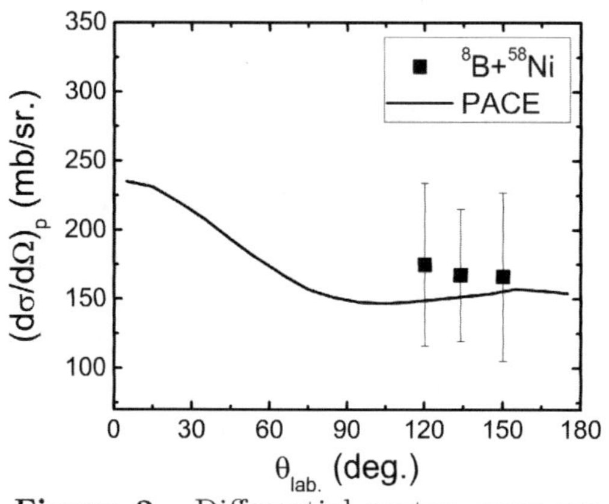

Figure 2. Differential proton cross section measured (symbol) and corresponding PACE [9] calculation (line) for E_{lab} = 30.0 MeV.

The experimental fusion excitation function, reported in Ref. [5], is represented by the filled circles in Fig. 3. The reported energies were determined iteratively and include a correction for mean energy loss in the target obtained by using successive Wong-model [14] fits to the data as weighting functions. Appropiate corrections were introduced to account for the respective isotopic composition when natural Ni targets were used. The repeat point at E_{fus} = 22.1 MeV, which was taken first using a ^{58}Ni target and later with a natural Ni target, validates the correction procedure. Two points were averaged to get the E_{fus} = 23.3 MeV point, leaving ten points in the graph. Also, in Fig. 3 the predictions of a barrier penetration model using the bare potential (solid curve), given by the São Paulo Potential (SPP) [15], are shown. With respect to these predictions, it can be seen that there is a strong *enhancement* of the fusion cross section below and above the respective Coulomb barrier for this proton-halo system. This behavior, which is different from the usual sub-barrier enhancement observed for many heavy-ion systems [7, 16–19], is possibly associated with possible proton-halo effects [5, 19], but an appropriate theoretical explanation is still due.

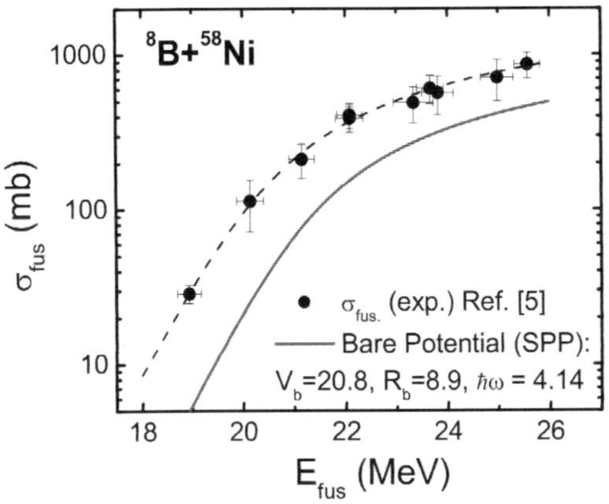

Figure 3. Fusion excitation function obtained in Ref. [5]. The dashed curve helps to guide the eye. The solid curve represents a one-dimensional BPM calculation using the barrier parameters obtained from the São Paulo Potential (SPP) [15].

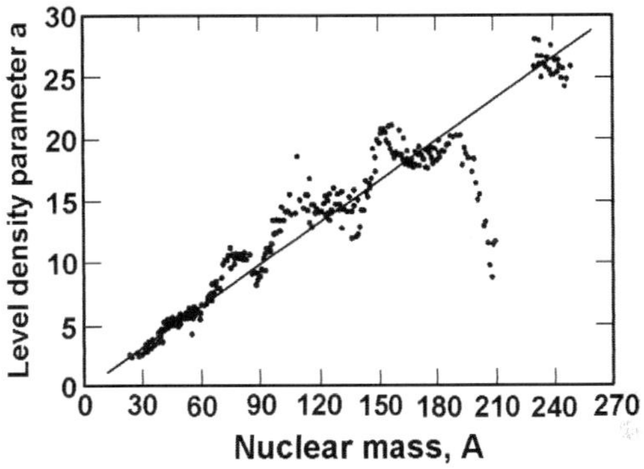

Figure 4. Values of the level density parameter a, as obtained from an analysis of 265 nuclei. The straight line corresponds to $a = A/9$ and approximates the average trend. Figure taken from Ref. [20].

Summarizing, in this section the most important facts regarding the determination of σ_{fus} were given. While σ_p is an experimental quantity which is essentially model-independent, one might wonder about a possible model-dependency introduced by the mapping $\sigma_p \rightarrow \sigma_{fus}$. In the following section the reliability of the respective procedure is investigated both, by studying the effect of variations in the input parameters to `PACE` [9] and by computing the M_p values with the alternate code `LILITA` [10, 11].

3. Level density effects and PACE vs LILITA

In general, both `PACE` and `LILITA` are Monte Carlo codes. The most important difference between the two codes lies in the way the transmission coefficients (T_l) are obtained. In `PACE`, the transmission coefficients are calculated for the compound-nucleus values of A and Z, and an extrapolation is made for subsequent decays [9]. On the other hand, in the present version of `LILITA`, an explicit calculation of all necessary transmission coefficients is made [11].

In Ref. [5], default values were used for most input parameters in `PACE` [9]. More specifically, the Yrast line was always determined by the liquid drop rotational energy, the A. J. Sierk fission barrier was assumed throughout the calculations, the level density parameter was $a = A/7.5$ (which is the value suggested by the author [9]), and the regular Wapstra mass table supplied with the code was used for all involved nuclei. In addition, the experimental fusion cross sections were used as an input (in an iterative way), which means that the code internally shifts the respective optical model transmission coefficients to reproduce these values.

The level density is one of the crucial ingredients of any statistical model calculation. This is given to the code through the level density parameter a. In Fig. 4, values of this level density parameter obtained from the analysis of 265 nuclei are shown [20]. The rapid decrease of a near shell closures is evident.

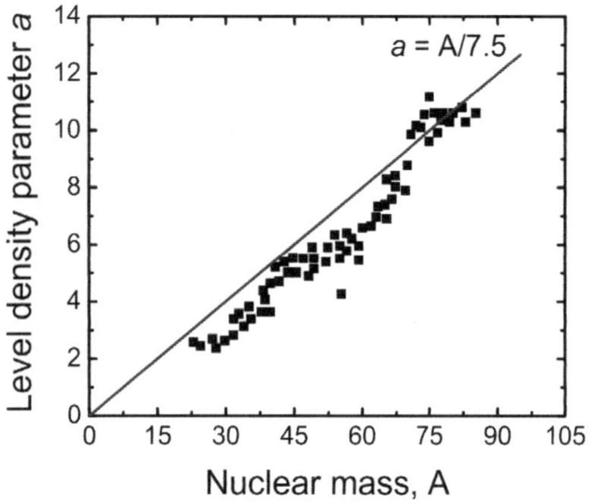

Figure 5. A zoom of the previous figure in the region A <90. The straight line corresponds to $a = A/7.5$ and approximates the average trend in the region 70< A <90.

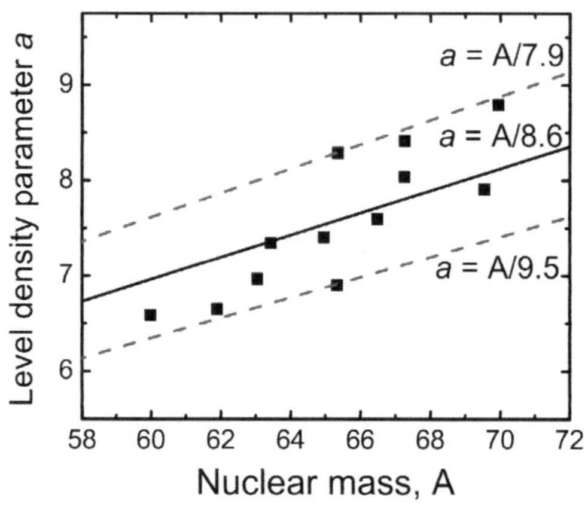

Figure 6. The level density parameter a plotted as a function of atomic mass number for 60< A <70. The straight line with $a = A/8.6$ approximates the average trend in this region, with upper (lower) limit given by $a = A/7.9$ ($a = A/9.5$), respectively.

In Fig. 5, the curve of the level density parameter given by $a = A/7.5$, corresponding to the default value in PACE [9], is plotted. This curve approximates the average trend of nuclei with atomic mass number A between 70 and 90. However, the atomic mass number of the compound nucleus corresponding to the ^{8}B+^{58}Ni system is $A = 66$ (^{66}As), which is below the group of nuclei described by the curve $a = A/7.5$. Taking this into account, the first step was to investigate the effect on the predictions of M_p, given by the code PACE, by changing $a = A/7.5$ to $a = A/8.6$. The curve corresponding to $a = A/8.6$, plotted in Fig. 6, approximates the average trend for nuclei with A between 60 and 70, which is the region where the mass number of the compound nucleus and its residual nuclei after evaporation would lie. The second step was to determine the predictions of M_p, given by the code PACE, by varying a within extreme limits. The limits were determined for 60< A <70, with an upper limit given by $a = A/7.9$ and lower limit given by $a = A/9.5$, as shown by the dashed lines in Fig. 6. As the final step, the predictions of M_p with $a = A/8.6$ were determined with the code LILITA [11]. The input parameters for LILITA were chosen as to have a meaningful comparison with the respective PACE calculations.

4. Results and discussion

The predictions of M_p for each energy, given by the code PACE [9] by replacing $a = A/7.5$ with $A/8.6$, are listed in Table 1. The maximum shift in M_p occurs at the highest energy. Based on these results, an estimated systematic uncertainty of ~5% must be assigned to the σ_{fus} values reported in Ref. [5] because of this effect.

In Fig. 7, the fusion cross sections obtained by applying the above correction to the level density parameter a, are compared to the previous values. The circles are the experimental

Journal of Physics: Conference Series **492** (2014) 012003 doi:10.1088/1742-6596/492/1/012003

Table 1. Shifts in M_p predicted by PACE [9] when changing the level density parameter a as discussed in the text.

$E_{c.m.}$ (MeV)	M_p $A/7.5$	M_p $A/8.6$	ΔM_p (%) $A/7.5$ vs $A/8.6$
18.9	2.41	2.30	4.2
20.1	2.42	2.32	4.1
21.1	2.43	2.33	4.0
22.1	2.42	2.32	4.1
22.1	2.42	2.32	4.1
23.2	2.44	2.33	4.4
23.4	2.44	2.33	4.4
23.7	2.43	2.32	4.5
23.8	2.44	2.32	4.6
25.0	2.44	2.32	4.8
25.6	2.43	2.31	5.0

values of σ_{fus} reported in Ref. [5] while the open squares are the respective values obtained for $a = A/8.6$. It can be seen that after considering this correction, σ_{fus} would still be well within the uncertainty limits previously reported.

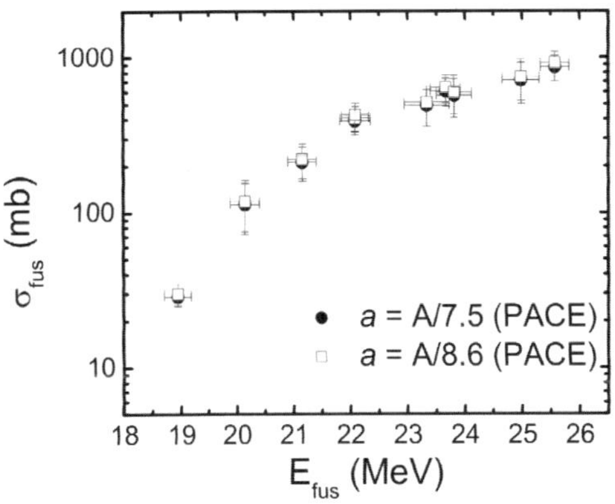

Figure 7. Fusion cross sections obtained for two values of the level density parameter a using PACE [9]. The black circles represent σ_{fus} for $a = A/7.5$ while the open squares refer to $a = A/8.6$.

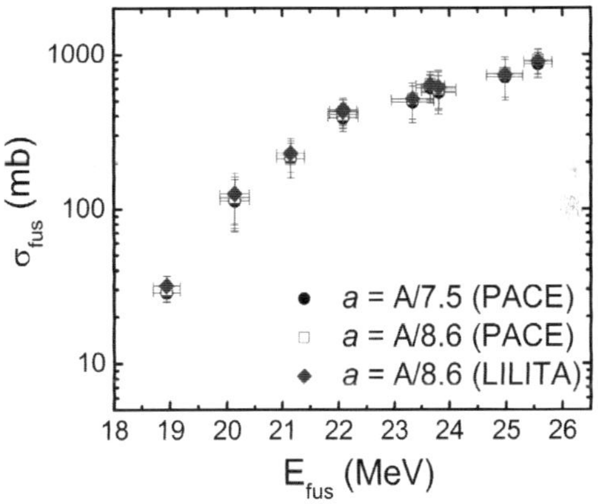

Figure 8. Sensitivity of fusion cross sections to the level density parameter a and to using different codes. The circles and the squares are the same as the ones plotted in Fig. 7. The diamonds are the σ_{fus} values obtained with the code LILITA [11] by using $a = A/8.6$.

On the other hand, when comparing the predictions of M_p (with code PACE [9]) for $a = A/8.6$ versus the ones obtained for the lower limit ($a = A/9.5$), a maximum systematic error of ~4% results. Similarly, by comparing the $a = A/8.6$ case versus the results obtained for the upper limit ($a = A/7.9$), a maximum systematic error of ~3% is obtained. By taking these and the previous results into account, properly convoluting the respective uncertainties, we conclude that the sensitivity of σ_{fus} to changes in the level density parameter a within extreme limits, can be accounted for by a total systematic uncertainty of ~6%.

Table 2. Shifts in M_p obtained by using the codes PACE [9] and LILITA [11] (for $a = A/8.6$).

$E_{c.m.}$ (MeV)	M_p PACE	M_p LILITA	$\Delta M_p(\%)$
18.9	2.30	2.21	4.0
20.1	2.32	2.22	4.2
21.1	2.33	2.25	3.3
22.1	2.32	2.26	2.9
22.1	2.32	2.26	2.7
23.2	2.33	2.35	0.6
23.4	2.33	2.36	1.4
23.7	2.32	2.33	0.5
23.8	2.32	2.31	0.5
25.0	2.32	2.35	1.0
25.6	2.31	2.40	4.0

In Table 2, the predictions of M_p with $a = A/8.6$ by using the codes PACE [9] and LILITA [11], are compared. A maximum shift in M_p of ~4% is observed. In Fig. 8, the diamonds are the respective σ_{fus} values obtained with LILITA. Once again, it is clear that after considering the M_p results obtained with the code LILITA, the respective σ_{fus} values would not show any important change. The effects of both, the sensitivity to changes in a (within PACE) and using an alternate code (LILITA), can be included in a global systematic uncertainty of ~7%.

5. Conclusions

The fusion cross sections (σ_{fus}) reported in Ref. [5] for the $^{8}B+^{58}Ni$ system, were reviewed to investigate possible model-dependency in the involved $\sigma_p \rightarrow \sigma_{fus}$ mapping. The key quantities in this mapping, the proton multiplicities (M_p), were previously calculated with the code PACE [9] by using default values for the respective input parameters. In the present work, the sensitivity of M_p to changes both in the involved level densities and in the relevant transmission coefficients, was investigated.

The maximum shift in M_p, obtained with the code PACE by changing the level density parameter within extreme limits, can be accounted for by assigning a systematic uncertainty of ~6% to the σ_{fus} values reported in Ref. [5].

On the other hand, a maximum shift of ~4% is observed in M_p when comparing with the results of the code LILITA [11], which, in contrast to PACE, explicitly calculates the necessary transmission coefficients. When applying the mentioned M_p shifts to the individual experimental points, the respective σ_{fus} values do not show any important change (Fig. 8).

The effects of both, the sensitivity to changes in a (within PACE) and using the alternate code (LILITA), can be included in a global systematic uncertainty of ~7%. This would not change the conclusions discussed in Ref. [5]. The reliability of the whole procedure was thus confirmed.

Acknowledgments
This work has been partially supported by CONACYT (México) and by the NSF (USA) under Grant No. 09-69456.

References
[1] Belyaeva T L, Aguilera E F, Martinez-Quiroz E, Moro A M and Kolata J J 2009 *Phys. Rev.* C **80** 064617
[2] Guimarães V *et al.* 2000 *Phys. Rev. Lett.* **84** 1862
[3] Kolata J J *et al.* 2001 *Phys. Rev.* C **63** 024616
[4] Aguilera E F *et al.* 2009 *Phys. Rev.* C **79** 021601(R)
[5] Aguilera E F *et al.* 2011 *Phys. Rev. Lett.* **107** 092701
[6] Canto L F, Gomes P R S, Donangelo R and Hussein M S 2006 *Phys. Rep.* **424** 1
[7] Keeley N, Raabe R, Alamanos N and Sida J L 2007 *Prog. Part. Nucl. Phys.* **59** 579
[8] Keeley N, Alamanos N, Kemper K and Rusek K 2009 *Prog. Part. Nucl. Phys.* **63** 396
[9] Gavron A 1980 *Phys. Rev.* C **21** 1
[10] Gomez del Campo J and Stokstad R G 1981 ORNL Report TM-7295
[11] Gomez del Campo J 2001 private communication.
[12] Lee M Y *et al.* 1999 *Nucl. Instrum. Methods Phys. Res.* A **422** 536
[13] Aguilera E F 2013 *Phys. Rev.* C **87** 014613
[14] Wong C Y 1973 *Phys. Rev. Lett.* **31** 766
[15] Chamon L C *et al.* 2002 *Phys. Rev.* C **66** 014610
[16] Vaz L C, Alexander J M and Satchler G R 1981 *Phys. Rep.* **69** 373
[17] Beckerman M 1983 *Rep. Prog. Phys.* **51** 1047
[18] Dasgupta M, Hinde D, Rowley N and Stefanini A 1998 *Annu. Rev. Nucl. Part. Sci.* **48** 401
[19] Aguilera E F and Kolata J J 2012 *Phys. Rev.* C **85** 014603
[20] Stokstad R G 1985 *Treatise on Heavy Ion Science* vol 3 ed D A Bromley (New York & London: Plenum) p 94

Journal of Physics: Conference Series **492** (2014) 012004 doi:10.1088/1742-6596/492/1/012004

Fusion, transfer and breakup of light weakly bound nuclei at near barrier energies.

P R S Gomes, J Lubian , J Rangel , D R O Otamar

Instituto de Física, Universidade Federal Fluminense, Av. Litoranea s/n, Gragoatá, Niterói, R.J., 24210-340, Brazil

Abstract. We discuss interesting questions and results concerned with reactions involving weakly bound nuclei at near barrier energies, particularly fusion, breakup, transfer and scattering. We concentrate in the presentation and discussion of recent results, and so we try to give a sort of state of art of this field at the present. PACS Numbers: 24.10Eq, 25.60Pj, 25.60Bx, 25.60Gc

1. Introduction

In the last decades, reactions between heavy íons at energies close to the Coulomb barrier have been a subject of great and renewed interest. Particularly, the fusion process has been extensively investigated. It is widely accepted that couplings to low lying collective states produce strong enhancement of the fusion cross section at sub-barrier energies, especially when nuclei with large static deformation, like ^{154}Sm, are involved [1, 2, 3, 4, 5, 6]. Transfer reactions can also be important doorway to enhance sub-barrier fusion cross sections [7, 8, 9, 10, 11, 12].

It has also been extensively investigated the elastic scattering between heavy ions at near barrier energies. It is well accepted that the elastic scattering presents a behaviour of the energy dependence of the interacting optical potential known as Threshold Anomaly (TA) [13, 14, 15], corresponding to a rapid variation of the real and imaginary parts of this potential when the energy decreases towards the Coulomb barrier energy. The real part of the potential shows a localized peak, while there is a sharp decrease of the imaginary part of this potential at energies close to the barrier. The decrease of the imaginary potential is due to the closure of the non-elastic channels at energies below the Coulomb barrier. There is a connection between the real and imaginary parts of the optical potential due to causality and subsequently they obey the dispersion relation [16, 17]. As consequence, the real potential shows an increase at those energies, in a bell shape form, when the imaginary potential decreases. The real potential can be written as $V_{eff} = V_0 + \Delta V$, where V_0 is the real potential at higher energies and ΔV is called polarization potential. So, it is found that the polarization potential is attractive, and it has the effect of enhancing the fusion cross section at sub-barrier energies, since it decreases the Coulomb barrier.

2. Reactions with weakly bound nuclei

The above situation may be different when weakly bound nuclei are involved in the reactions. These nuclei have low breakup energy threshold and the breakup feeds states in the continuum.

Following breakup, different processes may occur: non-capture breakup (NCBU), when neither fragment fuses, incomplete fusion (ICF), when part of the fragments fuse and sequential complete fusion (SCF), when all the fragments fuse. Total fusion (TF) is the sum of direct and sequential complete fusion (CF) and ICF. Reactions with weakly bound nuclei, both stable and radioactive (some of them with halo properties) have been intensively studied in the last years, and there are some comprehensive reviews on this subject up to 2007 [18, 19, 20]. In this paper we will concentrate on results which came out after those reviews, and so we will try to give a partial update of the status of art in this field, particularly based on works of ourselves and collaborators, but also on other reported important works.

One basic question in this field is whether the breakup process enhances or hinders the fusion cross section. However, before one tries to answer this question, one should be clear about what one is talking about. Also one has to know whether the investigation is concerned with CF or TF. Other important questions are: What are the effects on different energy regimes and on different target mass regions? Does the breakup affect the elastic scattering and the usual threshold anomaly of the optical potential at near barrier energies? Do different breakup threshold energies affect the total reaction cross section significantly? How large is the σ_{NCBU} compared with σ_{CF} or σ_{CF+ICF} ? How does it depend on the energy region and target mass? Is $\sigma_{transfer}$ important for most weakly bound projectiles? Experimental aspects to be considered are, among others: What is it measured: σ_{CF} or σ_{TF}= CF + ICF? Can ICF be separated from transfer channels leading to the same compound nucleus? If not, instead of TF, one measures TF + transfer cross section? Recently it has been observed [21, 22] that breakup following transfer of nucleons is also an important process and may predominate over the direct breakup at sub-barrier energies. Also, transfer channels were shown to be very important, at least for reactions like neutron-halo nuclei [23, 24, 25, 26, 27].

Concerning the elastic scattering, the energy dependence of the optical potential when weakly bound nuclei are involved has been shown, in several works, to have a different behaviour as that for tightly bound systems, namely, the TA and the attractive polarization potential owing to the couplings to inelastic and transfer channels. This is because the breakup channel may be important even below the barrier and produces repulsive polarization potential. So, the imaginary potential does not decrease at the barrier energy. Indeed, it may increase. Consequently, the real potential decreases at this energy region and fusion is suppressed owing to this effect. This behaviour is called breakup threshold anomaly (BTA) [28, 29]. Of course, the imaginary potential must decrease and vanish at lower energies. A detailed investigation of this effect is rather difficult because one needs precise elastic scattering data at sub-barrier energies, where the scattering is almost of Rutherford type and almost insensitive to the nuclear potential. There are also several works where there is a simultaneous fit of fusion and elastic scattering data and the optical potential is divided in an inner part connected with fusion and a superficial part connected with direct processes like breakup and transfer. It is then observed a threshold anomaly behaviour for the fusion potential [30]. A BTA behaviour for the direct reaction potential is usually found.

Other aspects of reactions with weakly bound nuclei have been investigated in the last years, such as the effect of breakup on inelastic scattering [31], quasi-elastic barrier distributions [32] and the effect of breakup on total reaction cross sections by using two different methods [33, 34]. However, in the following we will concentrate on the investigation of breakup effects on the fusion cross section.

3. Static and dynamic effects of breakup on the fusion cross section and the importance of the choice of the bare interaction potential

There are two main approaches to investigate this subject. The first one compares fusion data with predictions from some theory. For sure, the conclusions are strongly model dependent, and so it is very important that a reliable bare potential must be used in the calculations. The difference between theoretical and experimental cross sections, $\Delta\sigma_F$, is attributed to the ingredients missing in the theory. The second approach compares fusion data for different systems, including tightly bound ones. In any case, there are two kinds of effects to be investigated. First, there are the static effects related with different barrier characteristics, when compared with those for similar tightly bound systems, since the longer tail of the nuclear density of the weakly bound nuclei, especially the halo ones, produces a more diffuse potential, with a lower barrier, and consequently enhances the fusion cross section. The second kind is the dynamical effects associated with the strong coupling between the elastic and the breakup channels. This is a much more difficult effect to be analyzed and, in our opinion, the most interesting one. If one uses double folding potentials with realistic densities of the colliding nuclei as the bare potential, the possible static effects of the weakly bound nuclei are already taken into account, and so the differences between data and calculations show only the dynamic effects of the channels not included in the calculations.

When one compares data with theory, the comparison may be of different kinds: with single channel with standard densities of the nuclei ($\Delta\sigma_F$ is due to static + dynamic effects); single channel with realistic densities ($\Delta\sigma_F$ comes from couplings to all channels, but the static effect is already taken into account); coupled channel calculations taking into account all bound channels ($\Delta\sigma_F$ comes from the coupling to the continuum); Continuum Discretized Coupled Channel (CDCC) calculations taking into account also transfer channels ($\Delta\sigma_F$ should vanish). However, at the present it is not available the latter kind of calculations. The use of different interaction potentials may lead to controversies like for fusion of ^{6}He + ^{209}Bi [35] and ^{6}He + ^{238}U [23] data. Recently it has been shown [36] that there are no incompatibility between the two sets of data, which were originally analysed using different models. So, the choice of the bare interacting potential plays a major role in the analysis of the fusion cross section behaviour.

When one investigates the effect of breakup on the fusion cross sections for weakly bound systems, one must start with a standard behaviour of the fusion cross section to which the data should be compared. In order to do so, it is important to be able to compare different systems in the same graphic, what requires a proper normalization method to take into account trivial factors like different sizes and Coulomb barriers of the systems. Recently, Canto et al. [37] have shown that the traditional method of dividing cross sections by $\pi\ R_B^2$ and center of mass energies by V_B, where R_B and V_B are the radius and height of the Coulomb barrier, does not fully eliminate these geometrical effects.

4. Fusion functions, the UFF benchmark curve and systematic of fusion cross section of weakly bound systems

Recently, Canto et al. [37] have proposed a reduction procedure that can eliminate static effects, by the introduction of dimensionless function $F(x)$, called Fusion Function, and energy variable x, defined as $F(x) = \frac{2E\sigma_F}{\hbar\omega R_B^2}$ and $x = \frac{E-VB}{\hbar\omega}$. Here, V_B, R_B and $\hbar\omega$ are the barrier height, radius and the curvature, appearing in the parabolic approximation of the Coulomb barrier. For systems where channel coupling effects can be neglected and the fusion cross section can be approximated by Wong's formula [38], $F(x)$ becomes system independent and can be written as a Universal Fusion Function (UFF) written as $F_0(x) = ln[1+exp(2\pi x)]$. UFF is then used as the benchmark for comparisons with fusion data $F_{exp}(x)$. However, if one compares experimental fusion functions with the UFF, the differences would be due to the ingredients not considered in the calculations, such as all important low lying collective couplings, and not just breakup plus transfer. Also, the Wong model is known for not being valid at sub-barrier energies for light systems, as it is the situation of some those involving the above mentioned weakly bound projectiles. By these reasons, Canto et al. [37] proposed to renormalize the experimental fusion functions to take into account the possible failure of the Wong model and the effects of inelastic couplings. In the coupled channel calculations to be performed in this renormalization, a reliable bare potential must then be used. These renormalized fusion functions are then compared with UFF. Now, after this renormalization, the differences are dynamic effects due to the channels left out of the coupled channel calculations, in this case, breakup and transfer reactions. In their coupled channel calculations, the double folding Sao Paulo potential (SPP) was used [39]. This potential has been widely and successfully used in the last years, for a large variety of systems and energy ranges [40, 41], including fusion barrier distributions for weakly bound nuclei and fusion of halo nuclei [42, 43].

Complete and total fusion for tens of systems, both tightly and weakly bound (stable and neutron-halo), were compared with UFF and then a systematic behaviour could be observed [37, 44, 45, 46]. In figure 1 we show, as example, the comparison of the renormalized experimental fusion function for some representative systems. Data are from Ref. [23, 24, 25, ?, 12, 48, 49, 50, 51, 52]. We will discuss the results for fusion of the proton-halo ^{8}B [51, 52] in the next section.The linear scale is better to observe the behaviour at energies above the barrier, whereas the log scale is more suitable for sub-barrier energies. The reached systematic shows that complete fusion cross sections for stable weakly bound systems and total fusion cross sections for neutron-halo systems are around 30% suppressed at above barrier energies, when compared with UFF. So, the combined effect of breakup and transfer channels is to suppress fusion above the barrier. For total fusion of stable weakly bound system and for the ^{17}F + ^{208}Pb system [47], there is neither suppression nor enhancement above the barrier. Very recently, Fang et al. [53] were able to disentangle total fusion induced by ^{9}Be from transfer of one-neutron, and only if one adds both mechanisms the total fusion coincides with the UFF. Maybe in some works where total fusion was reported to have been measured, actually what was measured was the sum of total fusion plus transfer channels. At sub-barrier energies, all systems show some enhancement, although they are not as large as for low-lying excitations of deformed nuclei.

One may ask whether all systems investigated obey the systematic behaviour mentioned above. The answer is no. Among the tens fusion functions investigated from the available data in literature, only very few of them do not follow the systematic. Possible explanations for these anomalous behaviours might be either something very special with those systems or something wrong with the data or with the standard coupled channel calculations. Actually, some of those systems were measured again by other groups and/or experimental methods and the new measurements give results compatible with the systematic.

The puzzle concerning the explanation why the complete fusion is suppressed by the breakup at above barrier energies and enhanced at sub-barrier energies has been investigated recently by Gomes et al [54], in terms of polarization potentials. They have shown that at energies above the barrier, the direct breakup which produces repulsive polarization potential predominates over the attractive polarization potential originated by transfer channels, whereas at sub-barrier energies, where the breakup following transfer predominates over the direct breakup, a net attractive polarization potential which enhances the fusion cross section occurs. Indeed, recent calculations [55, 56, 57] show that the direct breakup produces repulsive polarization potentials, owing to the couplings among continuum breakup states (continuum-continuum couplings), whereas the breakup following transfer produces attractive polarization potential [54]. Those explanations are in agreement with experimental results for sub-barrier breakup [21, 22, 58]. Those results are also consistent with the polarization potentials derived from elastic scattering at lower energies, as can be observed in all extrapolations to low energies of the real polarization potential in several works. In those works usually there are no data at deeper sub-barrier energies, since the elastic scattering can hardly be distinguished from Rutherford scattering. However, in every case it is obvious that the imaginary part of the energy dependent optical potential has to vanish at very low energies, since all reaction channels will be closed at energies too much below the Coulomb barrier, So, due to the dispersion relation, the real part of the

Journal of Physics: Conference Series **492** (2014) 012004 doi:10.1088/1742-6596/492/1/012004

potential has to show the typical bell shape maximum at this energy range, corresponding to an attractive polarization potential which enhances the fusion.

5. Recent measurements of fusion cross sections for the proton-halo ^{8}B nucleus

Very recently, the first fusion cross section involving proton-halo nucleus, ^{8}B, was measured [51]. The results were very unexpected, since they show enhancement of the fusion cross section at energies above the barrier in relation with theoretical predictions which do not have any free parameter and with the UFF [59], as shown in figure 1. When one fits the data with free parameters [51] or in a simultaneous fit of fusion and elastic scattering data [60], the potential required to fit the data have an unusual long range. This leads to the conclusion that there is a special effect of proton-halo nuclei, different from those of neutron-halo nuclei, on the fusion cross section. This is particularly unexpected, because there are several works on the breakup of the ^{8}B on the presence of the ^{58}Ni target [61, 62, 63, 64] which show that the polarization potential produced by the breakup of ^{8}B is repulsive, what should lead to the suppression of fusion cross section. Also, the work on the elastic scattering for this systems, performed by Gomez Camacho et al. [65] shows the presence of the BTA, compatible with repulsive polarization potential and suppression of the fusion cross section above the barrier. However, it just came out a new paper reporting the fusion of ^{8}B with ^{28}Si [?], and the results agree very well with the systematic. So, more data on other proton-halo fusion are required to further investigate the possible existence of special proton-halo effect on the fusion or some problems with the derivation of fusion cross section for one of the mentioned systems.

6. Systematic of breakup and its effect on fusion, as a function of the charge of the target.

It is usually accepted that the Coulomb breakup increases with the charge of the target, and so it predominates over the nuclear breakup for heavy systems. Consequently, the breakup effects on the fusion cross section, such as the suppression of complete fusion at energies above the barrier, should be more important for heavier targets than for the light ones. This subject has already been investigated, but a systematic behaviour for the dependence of the complete fusion suppression as a function of the target mass, for the same projectile, has not yet reached [66, 67].

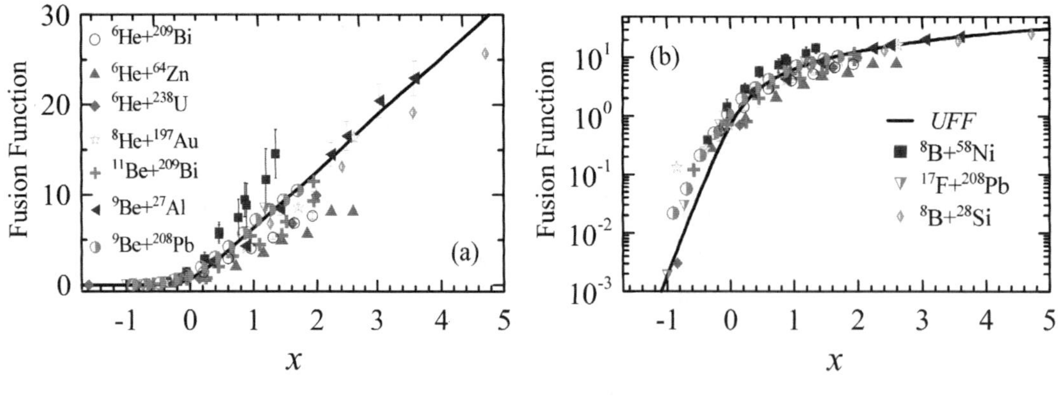

Figure 1. (color on line) Experimental renormalised fusion functions for total fusion of several weakly bound, proton-halo and neutron-halo systems. The curve is the universal fusion function. x is a reduced energy parameter. For details see the text and Ref. [37].

Journal of Physics: Conference Series **492** (2014) 012004 doi:10.1088/1742-6596/492/1/012004

Very recently, Otomar et al. [68] investigated the dependence of Coulomb, nuclear and total breakups of ^{6}Li on different targets, by performing CDCC calculations, which agree with the elastic scattering of those systems. It was shown that the nuclear breakup increases linearly with $A_T^{1/3}$ whereas the Coulomb breakup increases linearly with Z_T. However, a very interesting result of that work is that the Coulomb and nuclear breakups interfere destructively in such intense way that the total breakup is smaller than the Coulomb breakup for some energies. This result is shown in figure 2 for the ^{144}Sm and ^{208}Pb targets. In the same work [68], it was shown that the total breakup cross section is much smaller than the fusion cross section above and below the Coulomb barrier for light targets, like ^{59}Co, but that for heavier targets, like ^{144}Sm and ^{208}Pb, the breakup cross section is much larger than the fusion cross section at sub-barrier energies.

7. Systematic of total reaction cross sections for halo, weakly bound and tightly bound systems.

From the fits of elastic scattering angular distribution data, total reaction cross sections can be derived. A possible systematic behaviour for the total reaction cross section of weakly bound nuclei has been investigated in the last years, involving tens of systems (tightly bound, weakly bound and halo projectiles on targets with mass varying from light to heavy ones), and reported in a large amount of works. All of those papers use one or both of the presently two reduction methods used to compare different systems in the same graphic: the one proposed in 2005 by Gomes at al. [33] and the one proposed more recently by Shorto et al. [34], as an extension of the fusion function proposed by Canto et al.[37]. Both methods lead to qualitative similar results: the total reaction cross sections induced by halo nuclei are larger than for stable or non-halo weakly bound nuclei, which are similar or slightly larger than for tightly bound nuclei. A comprehensive work on this systematic was very recently published [69], using the latter method.

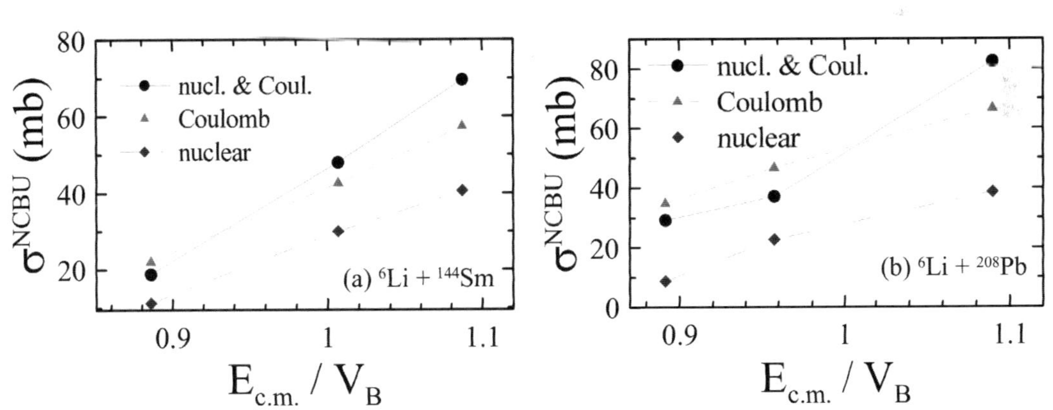

Figure 2. (color on line) Contribution of the Coulomb, nuclear and combined interactions to the NCBU cross section for the ^{6}Li + ^{144}Sm,^{208}Pb systems at near barrier energies. For details see the text and Ref. [68].

8. Conclusions

In conclusion, we can see that this fascinating subject is very alive, with a lot of interesting results coming out very recently. Several questions have already been answered, a lot has been learned, but there are still much to be done and learned, both experimentally and theoretically.

Acknowledgments
The authors would like to thank CNPq, CAPES, FAPERJ and PRONEX for the partial financial support.

References

[1] Stokstad R G *et al.* 1978 *Phys. Rev. Lett.* **41** 465; 1980 *Phys. Rev. C* **39** 2427
[2] Gil S *et al.* 1990 *Phys. Rev. Lett.* **65** 3100
[3] Gomes P R S *et al.* 1994 *Phys. Rev. C* **49** 245
[4] Leigh J R *et al.* 1995 *Phys. Rev. C* **52** 3151
[5] DiGregorio D *et al.* 1986 *Phys. Lett. B* **176** 322; 1989 *Phys. Rev. C* **39** 516
[6] Sargsyan V V *et al.* 2012 *Phys. Rev. C* **85** 037602; 2012 *Phys. Rev. C* **85** 01703; 2011 *Phys. Rev. C* **84** 064614
[7] Dasso C H *et al.* 1983 *Nucl. Phys. A* **405** 381; 1983 *Nucl. Phys. A* **407** 221; 1987 *Phys. Lett. B* **183** 141
[8] Beckerman M *et al.* 1980 *Phys. Rev. Lett.* **45** 1472; 1988 *Rep. Prog. Phys.* **51** 1047
[9] Gomes P R S *et al.* 1993 *Rev. Mex. Fis.* **39** 97
[10] Corradi L *et al.* 1990 *Z. Phys. A* **335** 55
[11] Gomes P R S *et al.* 1997 *J. Phys. G* **23** 1315
[12] Rehm K E 1991 *Annu. Rev. Nucl. Part. Sci* **41** 429
[13] Nagarajan M A, Mahaux C C and Satchler G R 1985 *Phys. Rev. Lett.* **54** 1136
[14] Satchler G R 1991 *Phys. Rep.* **199** 147
[15] Brandan M E and Satchler G R 1997 *Phys. Rep.* **285** 143
[16] Mahaux C, Ngo H and Satchler G R 1986 *Nucl. Phys. A* **449** 354
[17] Satchler G R and Love W 1979 *Phys. Rep.* **55** 183
[18] Canto L F , Gomes P R S. Gomes, Donangelo R. and Hussein M S 2006 *Phys. Rep.* **424** 1
[19] Liang J F , Signorini C 2005 *Int. J. Mod. Phys. E* **14** 1121
[20] Keeley N, Raabe R, Alamanos N , Sida J L 2007 *Prog. Part. Nucl. Sci.* **59** 579
[21] Luong D H *et al.* 2011 *Phys. Lett. B* **695** 105
[22] Rafiei R *et al.* 2010 *Phys. Rev. C* **81** 024601
[23] Raabe R *et al.* 2004 *Nature* **431** 823
[24] Lemasson A *et al.* 2009 *Phys. Rev. Lett.* **103** 232701
[25] Di Pietro A *et al.* 2004 *Phys. Rev. C* **69** 044613
[26] Kolata J J 2005 *Phys. Rev. C* **71** 067603
[27] DeYoung P A *et al.* 2005 *Phys. Rev. C* **71** 051601 (R)
[28] Gomes P R S *et al.* 2005 *J. Phys. G* **31** S1669
[29] Hussein M S, Gomes P R S, Lubian J, Chamon L C 2006 *Phys. Rev. C* **73** 044610
[30] Gomez Camacho A *et al.* 2010 *Phys. Rev. C* **82** 014616; 20210 *Phys. Rev. C* **82** 067601; 2008 *Phys. Rev. C* **77** 054606; 2007 *Phys. Rev. C* **76** 044609; 2008 *Rev. Mex. Fis.* **54** 56
[31] Woodard A E *et al. Nucl. Phys. A* **873** 17
[32] Monteiro D S *et al.* 2009 *Phys. Rev. C* **79** 014601
[33] Gomes P R S, Lubian J, Padron I , Anjos R M 2005 *Phys. Rev. C* **71** 017601
[34] Shorto J M B *et al.*2009 *Phys. Lett. B* **678** 77
[35] Kolata J J *et al.* 1998 *Phys. Rev. Lett.* **81** 4580
[36] Gomes P R S *et al.* 2011 *Phys. Lett. B* **695** 320
[37] Canto L F *et al.* 2009 *J. Phys. G* **36** 015109; 2009 *Nucl. Phys A* **821** 51
[38] Wong C Y 1973 *Phys. Rev. Lett.* **31** 766
[39] Chamon L C *et al.* 1997 *Phys. Rev. Lett.* **79** 5218 ; 2002 *Phys. Rev. C* **66** 014610
[40] Alves J J S *et al.* 2005 *Nucl. Phys. A* **748** 59
[41] Gasques L R, Chamon L C, Gomes P R S, Lubian J 2006 *Nucl. Phys. A* **764** 135
[42] Crema E , Chamon L C , Gomes P R S, 2005 *Phys. Rev. C* **72** 034610
[43] Crema E, Gomes P R S, Chamon L C 2007 *Phys. Rev. C* **75** 037601
[44] Gomes P R S, Lubian J , Canto L F 2009 *Phys. Rev. C* **79** 027606
[45] Gomes P R S *et al.* 2010 *Nucl. Phys. A* **834** 151c
[46] Lubian J *et al.* 2009 *Rev. Mex. Fis.* **55** 51
[47] Rehm K E *et al.* 1998 *Phys. Rev. Lett.* **81** 3341

[48] Signorini C *et al.* 1999 *Eur. Phys. J. A* **5** 7
[49] Dasgupta M *et al.* 1999 *Phys. Rev. Lett.* **82** 1395; 2004 *Phys. Rev. C* **70** 024606
[50] Marti G V *et al.* 2005 *Phys. Rev. C* **71** 027602
[51] Aguilera E F *et al* 2011 *Phys. Rev. Lett.* **107** 092701
[52] Pakou A *et al.* 2013 *Phys. Rev. C* **87** 014619
[53] Fang Y D *et al.* 2013 *Phys. Rev. C* **87** 024604
[54] Gomes P R S *et al.* 2012 *J. Phys. G* **39** 115103
[55] Santra S *et al.* 2011 *Phys. Rev. C* **83** 034616
[56] Canto L F *et al.* 2009 *Phys. Rev. C* **80** 047601
[57] Lubian J *et al.* 2007 *Nucl. Phys. A* **791** 24
[58] Shrivastava A *et al.* 2006 *Phys. Lett. B* **633** 463
[59] Rangel J *et al.* submitted to publication
[60] Gomez Camacho A , Aguilera E F , Lubian J, Gomes P R S 2013 *J. Phys. G* **40** 035103
[61] Lubian J *et al* 2009 *Phys. Rev. C* **79** 064605
[62] Lubian L , Correa T , Gomes P R S, Canto L F , Hussein M S 2010 *Nucl. Phys. A* **834** 802c
[63] Lubian J , Correa T , Gomes P R S, Canto L F 2008 *Phys. Rev.C* **78** 064615
[64] Lubian J , Nunes F M 2007 *J. Phys. G* **34** 513
[65] Gomez Camacho A *et al.* 2011 *Phys. Rev. C* **84** 034615; 2010 *Nucl. Phys A* **833** 156
[66] Gomes P R S *et al.* 2011 *Phys. Rev. C* **84** 014615
[67] Sargsyan V V, Adamian G G, Antonenko N V, Scheid W, Zhang H Q 2012 *Phys. Rev. C* **86** 054610
[68] Otomar D R, Gomes P R S, Lubian J, Canto L F, Hussein M S 2013 *Phys. Rev. C* **87** 014615
[69] Yang X P, Zhang G L, Zhang H Q 2013 *Phys. Rev. C* **87** 014603

Journal of Physics: Conference Series **492** (2014) 012005 doi:10.1088/1742-6596/492/1/012005

Optical model parallel description of elastic, fusion and breakup cross sections for systems with weakly bound projectiles

A Gómez Camacho[1], P R S Gomes[2], J Lubian[2], E F Aguilera[1]

[1]Departamento de Aceleradores, Instituto Nacional de Investigaciones Nucleares, Apartado Postal 18-1027, C.P. 11801, México, D.F.

[2]Instituto de Fisica, Universidade Federal Fluminenese, Avenida Litoranea s/n, Gragoatá, Niterói, RJ, cep 24210-340, Brazil

E-mail: arturo.gomez@inin.gob.mx

Abstract. A brief description is presented of the results obtained in recent years for the simultaneous analysis of elastic and fusion cross section data of nuclear reactions for several nuclear systems with weakly bound and halo projectiles. The method used in this description, consists of simultaneously determine the parameters of fusion U_F and direct reaction U_{DR} polarization potentials of Woods-Saxon geometric shapes, that fit the elastic and fusion data. As a matter of fact, U_F is an energy dependent potential, with real V_F and imaginary W_F components, that is responsible for fusion reactions. Similarly, U_{DR} is also energy dependent with real V_{DR} and imaginary W_{DR} parts, that accounts for direct reactions. A general finding for all the systems presented is that, the real and imaginary parts of the fusion potential and direct reaction potentials, are related by a dispersion relation and their energy dependence around and below the Coulomb barrier, show the so-called Breakup Threshold Anomaly. The effect of breakup reactions on fusion cross sections is studied by analyzing the separate effect of the absorption potential W_{DR} and the fusion barrier rising produced by V_{DR}.

1. Introduction

Nuclear reactions involving weakly bound and halo projectiles have been the source of extensive investigations in recent years[1]. Since this type of projectiles have low threshold energies, thus can be split into different fragments even at low collision energies, hence large breakup and transfer yields have been measured at energies around and below the Coulomb barrier. Similarly, due to the appreciable breakup cross sections, other nuclear reaction mechanisms like elastic scattering and fusion, are strongly affected by this process[1, 2, 3]. As a matter of fact, several nuclear processes are present in reactions with weakly bound and halo nuclei, besides elastic scattering and inelastic scattering, fusion can occur in several forms, i.e., a) Direct Complete Fusion (DCF), when the whole weakly bound projectile fuses to the target without splitting, b) Sequential Complete Fusion (SCF) when the projectile splits into different fragments but all of these fuse to the target, c) Incomplete Fusion (ICF), when after the splitting, only some of the projectile fragments are captured by the target and others feed the continuum d) Non-Capture Breakup NCBU when all the fragments feed the continuum, i.e., none is captured and e) Nucleon Transfer when some neutrons or protons are transferred to the target. Experimentally, it is very difficult to distinguish between the CF and SCF processes, so these are normally reported as complete fusion. Total fusion is composed of all the fusion mechanisms, $\sigma_{TF} = \sigma_{DCF} + \sigma_{SCF} + \sigma_{ICF}$. Very few experimental works have been performed that report exclusive CF and ICF cross section measurements[4, 5, 6, 7], in these cases complete fusion is considered as the fusion of the total charge of the projectile and the target. Basically, the small threshold energy associated to the breakup process influences fusion in two different manners. First, through the *static effect* connected to the $l = 0$ Coulomb barrier, that is, the height and curvature of the Coulomb barrier might enhance or hinder fusion. Secondly, through the *dynamical effect*, associated to the strong couplings to different reaction channels, that produce polarization potentials that can be attractive or repulsive. These polarization potentials, in turn, can lower or rise the fusion barrier and consequently enhance or hinder the fusion cross section. Normally, only the combined static and dynamical effects on fusion has been investigated, however, recently it has been possible to separate the particular effect for weakly bound systems[8, 9].

In several recent works, the effect of breakup on fusion has been studied by determining the dynamical polarization potential that arises from couplings between the incident elastic channel to different reaction channels. This polarization potential $U(r,E)$ has been calculated by a simultaneous fitting to elastic and fusion cross section data for various weakly bound systems[10, 11, 12, 13, 14, 15, 16, 17]. As a matter of fact, in order to do the simultaneous fitting, the energy dependent polarization potential $U(r,E)$ is split into a fusion part $U_F(r,E)$ responsible to fusion couplings and a direct reaction part $U_{DR}(r,E)$ related to any other direct reaction couplings. In turn, the fusion and direct reaction polarization potentials have real and imaginary potentials, $V_F(r,E)$, $W_F(r,E)$ and $V_{DR}(r,E)$, $W_{DR}(r,E)$ respectively, that are correlated by the dispersion relation[18, 19], i. e.,

$$V_{F,DR}(R_{sa},E) = \frac{1}{\pi}\mathcal{P}\int_0^{\infty} \frac{W_{F,DR}(R_{sa},E')}{E'-E}dE', \tag{1}$$

where R_{sa} is the strong absorption radius. This relation establishes a constraint on the energy dependence of $V_{F,DR}$ and $W_{F,DR}$ leading to sharp variations of these functions around the Coulomb barrier V_B. For tightly bound systems, the polarization potentials $V_F(E)$ and $V_{DR}(E)$, that arise from couplings to fusion and bound state excitation channels, become attractive. This is due to a decreasing behavior of $W_F(E)$ and $W_{DR}(E)$ as the energy decreases below the Coulomb barrier. That is, as the energy decreases, fusion and direct reaction channels become eventually closed, an effect manifested by decreasing functions $W_F(E)$ and $W_{DR}(E)$. Consequently, this is accompanied by an increase attraction represented by $V_F(E)$ and $V_{DR}(E)$. This conjugated energy behavior between the real and imaginary parts of the polarization potentials has been termed as Threshold Anomaly (TA)[18, 19, 20] and is a direct consequence of the causality principle. An important immediate effect of the dynamical attractive potentials $V_F(E)$ and $V_{DR}(E)$, is to lower the nominal Coulomb barrier, that is, the barrier V_B resulting from the static nuclear bare V_{bare} and Coulomb V_{Coul} potentials. Hence, fusion is enhanced at energies around and below the barrier energy.

A different situation happens when weakly bound nuclei are considered. In these cases the breakup channel remains open well below the Coulomb barrier. The direct reaction polarization potential should account for appreciable couplings to channels in the continuum even at low energies. Then, the absorptive direct reaction W_{DR} can not decrease since important breakup reactions occur around and below V_B. Hence, from the dispersion relation, the real polarization potential V_{DR} that arises from couplings to these continuum states results repulsive. On the other hand, fusion cross sections become hindered, a fact that is accounted for, by a decreasing W_F (fusion channel progressively closes as the energy decreases towards the barrier V_B), which is accompanied by an attractive V_F. The effect of non-negligible breakup reactions at low energies on fusion cross sections is twofold, namely, the incident flux that is absorbed into breakup reactions which considered by W_{DR} and the rising of the fusion barrier due to V_{DR}. This particular energy dependent behavior, in contraposition to reactions with tightly bound systems, is known as Breakup Threshold Anomaly (BTA)[21, 22, 23]. The BTA has been studied in a variety of systems involving weakly bound and halo projectiles with targets of different charges and masses. For instance ^{9}Be with ^{27}Al[24], ^{64}Zn[23, 25], ^{144}Sm[26, 27], ^{208}Pb and ^{209}Bi[28, 29]. ^{6,7}Li in reactions with ^{208}Pb[21, 30], ^{27}Al[31, 32], ^{144}Sm[33], ^{58,64}Ni[34, 35], ^{59}Co[36], ^{90}Zr[37], ^{138}Ba[23, 38], ^{28}Si[39]. Halo systems such as ^{8}B with ^{58}Ni[14], and ^{6}He impinging on heavy targets ^{209}Bi[40] and ^{208}Pb[41].

In this paper, a brief description of the results obtained in recent years, of the simultaneous analysis of fusion and elastic scattering angular distribution data for several nuclear systems with weakly bound projectiles. The basic idea is to determine the parameters of the real and imaginary parts of the fusion U_F and direct reaction U_{DR} polarization potentials by a χ^2-analysis of the data. As will be described in the next sections, the calculated energy dependence of the real and imaginary parts of the fusion polarization potential V_F and W_F not only are linked by the dispersion relation, but also show a behavior consistent with the usual Threshold Anomaly. That is, the incident flux that is absorbed into the fusion channel, represented by $W_F(E)$ decreases as the energy also diminishes below the Coulomb barrier V_B. On the other hand, the associated virtual potential V_F, that arises from couplings between the incident elastic channel and the fusion ones, becomes attractive with a bell-shaped energy dependence centered around the barrier V_B. As a matter of fact, an important result of the present analysis is that, for all the systems studied, the calculated imaginary absorption potential $W_{DR}(E)$, which is the potential that accounts for the flux that is absorbed into direct reactions (basically breakup for weakly bound systems), does not decrease around and below the barrier, as does $W_F(E)$. Conversely, the determined dynamic direct reaction real polarization potential $V_{DR}(E)$ becomes repulsive around the barrier energy. Therefore, $V_{DR}(E)$ has the important effect of increasing the fusion barrier and thus hinders fusion. This particular energy dependent behavior of $V_{DR}(E)$ and $W_{DR}(E)$ is linked by the dispersion relation. Finally, the effect on fusion due to $V_{DR}(E)$, $W_{DR}(E)$ is investigated. First, the separate effect of either $V_{DR}(E)$ and $W_{DR}(E)$ is studied, then the combined effect is calculated. Since a detailed description of the methodology used in these calculations has been given in several papers, here only a very brief account is given. A complete description of the results of the calculations is given for some of the systems studied, but for other systems only the conclusions are mentioned.

2. Basic features of the model.

The total interaction between projectile and target is given by,

$$\mathcal{V}(r,E) = V_{Coul}(r) - V_{bare}(r) - U(r,E), \tag{2}$$

where V_{coul} is the repulsive Coulomb interaction between the nuclei, $V_{bare}(r)$ is the energy independent average nuclear potential and $U_(r,E)$ is the nuclear dynamic polarization potential that arises from couplings between the incident elastic channel to fusion and direct reaction ones. The polarization potential has real and imaginary parts, i.e.,

$$U(r,E) = V(r,E) + iW(r,E), \tag{3}$$

where, at the strong absorption radius R_{sa}, the energy dependent functions $V(E)$ and $W(E)$ are linked by the dispersion relation (Eq.(1)). The total absorption potential $W(r,E)$ can be split into fusion and direct reaction absorption parts,

$$W(r,E) = W_F(r,E) + W_{DR}(r,E). \tag{4}$$

A formal justification of why the total absorption potential W can be divided into different absorption channels is given in Refs.[11, 42, 43]. Consequently, the total real polarization potential can also be split into fusion and direct reaction parts,

$$V(r,E) = V_F(r,E) + V_{DR}(r,E), \tag{5}$$

where $V_F(r,E)$ and $W_F(r,E)$ as well as $V_{DR}(r,E)$ and $W_F(r,E)$ are related by a dispersion relation. Fusion and direct reaction cross sections are given by,

$$\sigma_{F,DR}(E) = \frac{2}{\hbar v} < \chi_a^{(+)} \, |W_{F,DR}(E)| \, \chi_a^{(+)} >, \tag{6}$$

where $\chi_a^{(+)}$ represents the relative motion wave function between the nuclei and v their relative velocity. Thus, the total reaction cross section is calculated by,

$$\sigma_R(E) = \sigma_F(E) + \sigma_{DR}(E). \tag{7}$$

The fusion potentials $V_F(r,E)$ and $W_F(r,E)$ are assumed of Woods-Saxon volume shapes, i.e,

$$V_F(r,E) = V_{0,F}(E) f_F(r) \;\; ; \;\; W_F(r,E) = W_{0,F}(E) f_F(r), \tag{8}$$

where,

$$f_i(r) = \frac{1}{1+\exp[(r-R_i)/a_i]}, \qquad i = F, DR, \tag{9}$$

$V_{F,0}(E)$ and $W_{F,0}(E)$ are respectively the strengths of $V_F(r,E)$ and $W_F(r,E)$, a_F the diffuseness and $R_F = r_F(A_p^{1/3} + A_T^{1/3})$, r_F being the reduced radius parameter. Similarly, the direct reaction potentials have surface Woods-Saxon shapes,

$$V_{DR}(r,E) = -4a_{DR}V_{0,DR}(E)\frac{df_{DR}}{dr}, \quad W_{DR}(r,E) = -4a_{DR}W_{0,DR}(E)\frac{df_{DR}}{dr}, \tag{10}$$

where $R_{DR} = r_{DR}(A_p^{1/3} + A_T^{1/3})$. $V_{0,DR}$ and $W_{0,DR}$ are the potential strengths, a_{DR} the diffuseness and r_{DR} the reduced radius parameters. The strengths, diffuseness and reduced radii of the potentials are fixed during the simultaneous fitting of elastic and fusion data. Clearly, the energy dependence of the potentials can be determined around the Coulomb barrier.

3. The Breakup Threshold Anomaly Investigation.

The energy dependence of the fusion and direct reaction dynamic polarization potentials has been studied for several systems with weakly bound projectiles[23-41]. Here, only a general discussion is given for two cases. Figures 1 and 2, show the results of the fusion and direct reaction dynamic polarization potentials for the systems $^{8}B+^{58}Ni$ and $^{6}Li+^{28}Si$. The values of $W_F(E)$ at $R_{sa} = 12.1$ fm, as obtained by the χ^2-analysis of elastic and fusion data for $^{8}B+^{58}Ni$, shown by the full circles in Fig.1b, decrease as the energy is lowered below the Coulomb barrier $V_B = 20.8$ MeV. This means that the fusion channel becomes progressively closed. On the other side, the real fusion polarization potential $V_F(E)$ as obtained by the fitting calculation and shown by the full circles of Fig.1a, has an increasing-decreasing behavior as the energy is lowered just around V_B and is attractive. The line that follows the behavior of $V_F(E)$ is the integration according to the dispersion relation of the linear functions given for $W_F(E)$(see details in Ref.[44]). So, $W_F(E)$ and $V_F(E)$ have an energy dependence consistent with the usual Threshold Anomaly as in the case of tightly bound systems. The obtained values for the direct reaction imaginary $W_{DR}(E)$ (squares in Fig.1b) are very stable at high energies but shows an

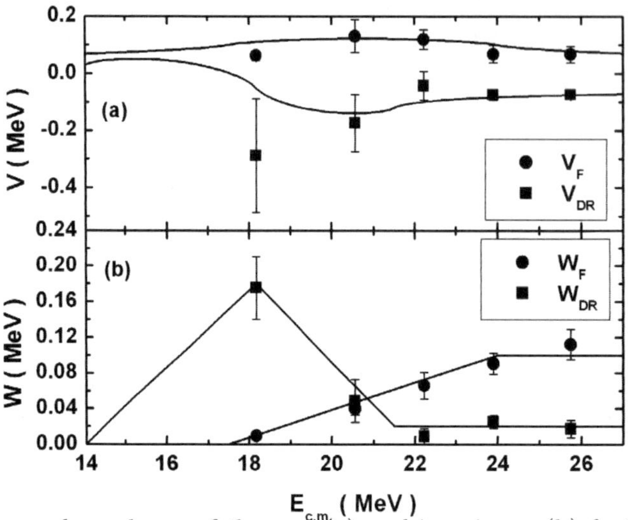

Figure 1. (a) Energy dependence of the real (a) and imaginary (b) fusion and direct reaction polarization potentials at the strong absorption radius R_{sa} for the system ^{8}B+^{58}Ni.

appreciable increase at the lowest energy. This is related to the fact that flux into breakup reactions continues to be important at low energies. The real potential $V_{DR}(E)$, that accounts for coupling effects to breakup channels, becomes repulsive and with increasing strength, as seen by the squares of Fig.1a. The corresponding line is that obtained by the dispersion relation using the parametrization assumed for $W_{DR}(E)$. Figure 2 shows the results for the system ^{6}Li+^{28}Si[15], where as before, $W_{DR}(E)$ increases just as the energy approaches the barrier, while $V_{DR}(E)$ is repulsive. Also, the fusion potentials show the usual threshold anomaly around the barrier energy. The energy dependence of the polarization potentials for these two systems, show that instead of the TA, the BTA is present. As is the case of many other weakly bound and halo systems investigated in recent years.

4. Effect of the direct reaction real polarization potential on the Coulomb barrier.

The effect on the nominal Coulomb barrier V_B, due to the calculated dynamical fusion V_F and direct reaction V_{DR}, real potentials, can be calculated from Eq.(2),

$$Re[\mathcal{V}(r,E)] = V_{Coul}(r) - V_{bare}(r) - V_F(r,E) - V_{DR}(r,E), \tag{11}$$

where $Re[\mathcal{V}(r,E)]$ is the real part of the potential $\mathcal{V}(r,E)$ of Eq.(2). So, the attractive or repulsive effects of $V_F(r,E)$ and $V_{DR}(r,E)$ on the nominal Coulomb barrier can be studied. In figure 3, the effects on the nominal Coulomb barrier (l=0) by the fusion V_F and direct reaction V_{DR} polarization potentials for the system ^{8}B+^{58}Ni at the lowest measured energy $E_{c.m.} = 18.17$ MeV is presented. The nominal Coulomb barrier (full line) is given by the Coulomb potential V_{coul} and the attractive average nuclear potential V_{bare}, which is determined at the highest energy where polarization effects are negligible. The barrier lowering effect of the attractive fusion potential $V_F(r)$ of Eq.(8) is shown by the dashed-line. However, when V_{DR} is considered (dotted-line), the total effect is a net barrier increase. Thus, the incident flux faces a higher barrier for fusion, hence it is suppressed. This fusion suppression is dynamic, i.e., varies with the energy as seen in Fig.4, where the barrier rising effect decreases as the energy increases.

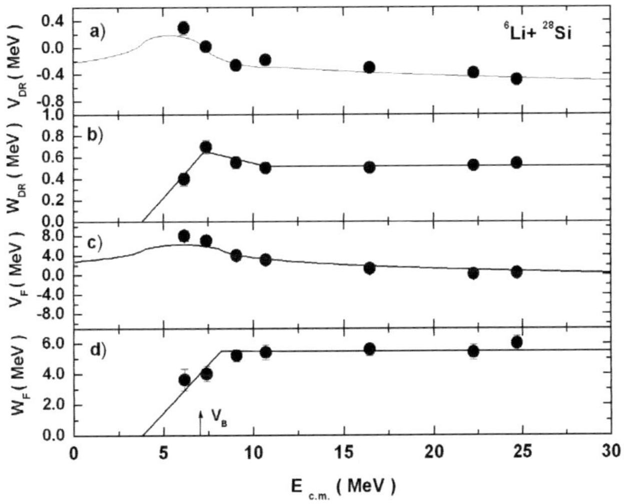

Figure 2. Energy dependence of the real(a) and imaginary (b) direct reaction potentials. Corresponding fusion potentials are shown in (c) and (d).

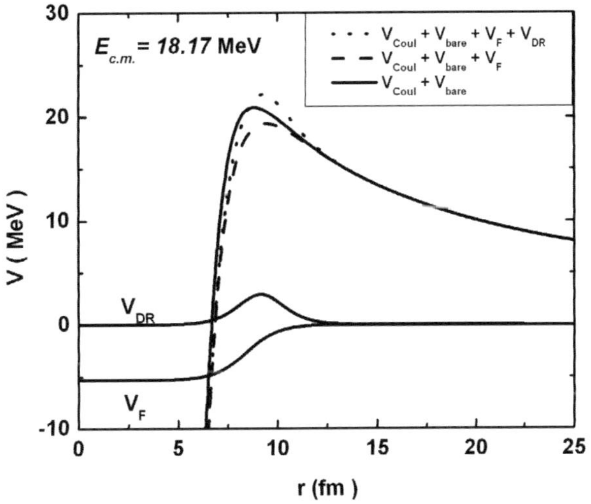

Figure 3. Effect of the fusion V_F and direct reaction V_{DR} polarization potentials on the nominal Coulomb barrier for the system ^{8}B+^{58}Ni at $E_{c.m.} = 18.17$ MeV.

5. Effect of the real and imaginary direct reaction polarization potential on the fusion cross section.

The particular effect on the fusion cross section due to the lowering and rising effects of $V_F(r, E)$ and $V_{DR}(r, E)$ respectively, can be calculated by,

$$R_{V_F,V_{DR}}(E) = \frac{\sigma_F(V_F(E),\ V_{DR}(E))}{\sigma_F(V_F = 0,\ V_{DR} = 0)}. \tag{12}$$

In the numerator, $\sigma_F(V_F(E),\ V_{DR}(E))$, two cases are considered, that when, $V_F(E) \neq 0$ but $V_{DR}(E) = 0$, and both $V_F(E) \neq 0$, $V_{DR}(E) \neq 0$. The denominator, $\sigma_F(V_F = 0, V_{DR} = 0)$ is the

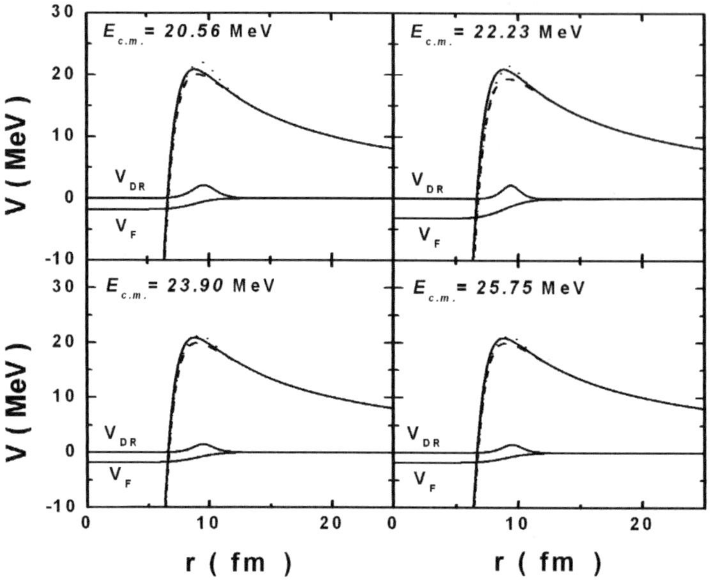

Figure 4. Same as Fig.3 but for the energies shown.

fusion cross section when both potentials are turned-off, that is, it represents the fusion cross section corresponding to the nominal barrier. $\sigma_F(V_F(E) \neq 0,\ V_{DR}(E) \neq 0)$ corresponds to the actual calculation for fusion cross section resulting from the simultaneous analysis. Fig.5a shows the results for the ^{8}B+^{58}Ni system. As seen, $R_{V_F,V_{DR}=0} > 1$ (full circles), i.e., fusion is enhanced by V_F. When V_{DR} is considered, $R_{V_F,V_{DR}} < 1$, thus the net effect is of fusion suppression.

Similarly, the effect of breakup reactions on fusion due to barrier rising produced by V_{DR} and loss of incident flux into direct reactions accounted for by W_{DR}, can be studied by,

$$R_{V_{DR},W_{DR}}(E) = \frac{\sigma_F(V_{DR}(E),\ W_{DR}(E))}{\sigma_F(V_{DR}=0,\ W_{DR}=0)}. \tag{13}$$

Fig.5b shows the results, where three cases are calculated, $V_{DR}(E) = 0$ and $W_{DR}(E) \neq 0$ (circles) ,$V_{DR}(E) \neq 0$ and $W_{DR}(E) = 0$ (squares), and finally $V_{DR}(E) \neq 0$ and $W_{DR}(E) \neq 0$ (stars). Both, V_{DR} and W_{DR} suppress fusion, but the effect is stronger for V_{DR} and for low energies.

6. Summary.

Important results obtained in recent works, regarding the simultaneous χ^2-analysis of elastic scattering angular distributions and fusion cross sections for weakly bound systems have been discussed. In particular, the cases ^{8}B+^{58}Ni and ^{6}Li+^{58}Ni have been presented in detail. However, similar conclusions are valid for other systems, for instance, the weakly bound ^{6}Li and ^{7}Li on targets ^{28}Si and ^{58}Ni, ^{9}Be on ^{208}Pb or the neutron halo ^{6}He on ^{209}Bi. It has been shown that the Breakup Threshold Anomaly appears for these systems. This is a direct consequence of a non-decreasing direct reaction imaginary potential $W_{DR}(E)$ (open channels) and the repulsive character of $V_{DR}(E)$ at energies around and below the barrier V_B. On the contrary, the fusion potentials $V_F(E)$ and $W_F(E)$ show the usual Threshold Anomaly. The effect on the fusion barrier due to the real parts $V_F(E)$ and $V_{DR}(E)$ has been studied, where a net fusion suppression is found, particularly at the lowest energies. Also, the effect on fusion by

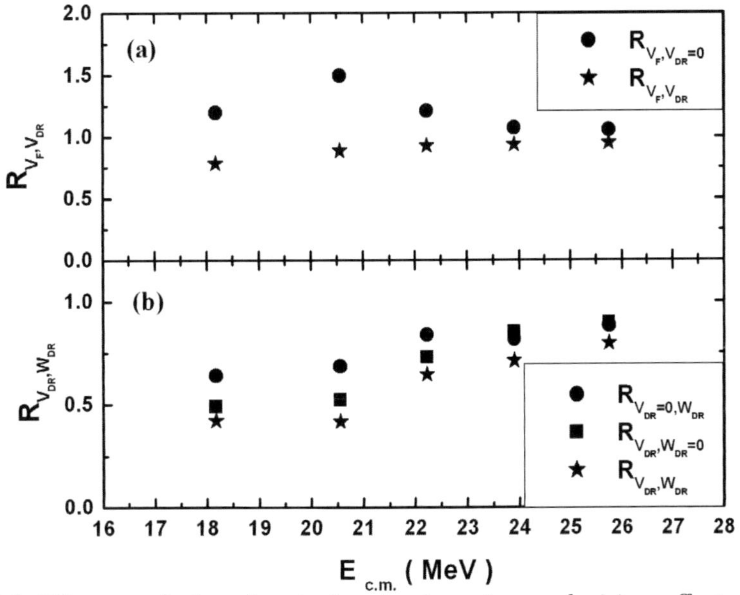

Figure 5. (a) Effect on fusion due to barrier lowering and rising effects produced by the attractive V_F and repulsive V_{DR} real polarization potentials. (b) Effect due to direct reactions (basically breakup), on fusion cross sections from barrier rising (V_{DR}) and loss of flux (W_{DR}).

breakup reactions due to loss of flux and barrier rising, accounted for by $W_{DR}(E)$ and $V_{DR}(E)$ respectively, has been calculated. It has been found that both of these potentials contribute to fusion suppression.

Acknowledgments A. Gómez Camacho, E.F. Aguilera thank the financial support of CONACYT, México. P.R.S. Gomes. and J. Lubian. acknowledge the financial support from CNPq, FAPERJ and PRONEX.

References

[1] Canto L F, Gomes P R S, Donangelo R, Hussein M S 2006 *Phys. Rep.* **424** 1
[2] Bertulani C A, Hussein M S and Munzenberg G, 2001 *Physics of Radioactive Beams* (Nova Science, New York)
[3] Bertulani C A, Canto L F, Hussein M S, 1993 *Phys. Rep.* **226** 281
[4] Dasgupta M et al., 2004 *Phys. Rev.* C **70** 024606
[5] Dasgupta M et al., 1999 *Phys. Rev. Lett.* **82** 1395
[6] Gomes P R S et al., 2006 *Phys. Rev.* C **73** 064606
[7] Gomes P R S et al., 2006 *Phys. Lett.* B **634** 356
[8] Canto L F, Gomes P R S, Lubian J, Chamon L C, Crema E, 2009 *J. Phys. G: Nucl. Part. Phys.* **36** 015109
[9] Canto L F, Gomes P R S, Lubian J, Chamon L C, Crema E, 2009 *Nucl. Phys.* A **821** 51
[10] Udagawa T, Tamura T, 1984 *Phys. Rev.* C **29** 1922
[11] Udagawa T, Kim B T, Tamura T, 1985 *Phys. Rev.* C **32** 124
[12] Gómez Camacho A, Gomes P R S, Lubian J, Padrón I, 2008 *Phys. Rev.* C **77** 054606
[13] Gomes P R S et al , 2009 *Nucl. Phys.* A **828** 233
[14] Gómez Camacho A, Aguilera E F, Gomes P R S, Lubian J, 2011 *Phys. Rev.* C **84** 034615
[15] Gómez Camacho A, Gomes P R S, Lubian J, 2010 *Phys. Rev.* C **82** 067601
[16] So W, Wong S W, Kim B T, Udagawa T, 2004 *Phys. Rev.* C **69** 064606
[17] So W, Wong S W, Kim B T, Udagawa T, 2005 *Phys. Rev.* C **72** 064602
[18] Satchler G R, 1991 *Phys. Rep.* **199** 147
[19] Mahaux C, Ngo H, Satchler G R, 1986 *Nucl. Phys.* A **449** 354

[20] Nagarajan M A, Mahaux C C, Satchler G R, 1985 *Phys. Rev. Lett.* **54** 1136
[21] Hussein M S, Gomes P R S, Lubian J, Chamon L C, 2006 *Phys. Rev.* C **73** 044610
[22] Hussein M S, Gomes P R S, Lubian J, Chamon L C, 2007 *Phys. Rev.* C **76** 019902(E)
[23] Gomes, P R S et al., 2005 *J. Phys.* G **31** S1669
[24] Gomes, P R S, et al., 2004 *Phys. Rev.* C **70** 054605
[25] Moraes S B et al., 2000 *Phys. Rev.* C **61** 064608
[26] Gómez Camacho A, Gomes P R S, Lubian J, Aguilera E F, Padrón I, 2007 *Phys. Rev.* C **76** 044609
[27] Gomes, P R S, et al., 2009 *Nucl. Phys.* A **828** 233
[28] Wooliscroft R J, et al., 2003 *Phys. Rev.* C **68** 014611
[29] Signorini C, et al., 2000 *Phys. Rev.* C **61** 061603
[30] Keeley N et al., 1994 *Nucl. Phys.* A **571** 326
[31] Figueira J M., *et al.* 2007 *Phys. Rev.* C **75** 017602. Figueira J M., *et al.* 2006 *Phys. Rev.* C **73** 054603
[32] Fernandez Niello J O. et al., 2007 *Nucl. Phys.* A **787** 484
[33] Figueira J M., *et al.* 2010 *Phys. Rev.* C **81** 024613
[34] Biswas M, et al., 2008 *Nucl. Phys.* A **802** 67
[35] Gómez Camacho A, et al., 2010 *Nucl. Phys.* A **833** 156
[36] Souza F A et al 2007 *Phys. Rev.* C **75** 044601
[37] Kamawat H et al 2008 *Phys. Rev.* C **78** 044617
[38] Maciel A M M et al 1999 *Phys. Rev.* C **59** 2103
[39] Pakou A et al 2004 *Phys. Rev.* C **69** 054602. Pakou A et al 2003 *Phys. Lett.* B **556** 21. Pakou A et al 2009 *Eur. Phys. J.* A **39** 187
[40] Garcia A R *et al* 2007 *Phys. Rev.* C **76** 067603
[41] Sanchez-Benitez A M *et al* 2008 *Nucl. Phys.* A **803** 30
[42] Feshbach H., 1967 *Ann. Phys.* (N.Y), **19** 287
[43] Kawai K., Kerman A.K. and McVoy K., 1973 *Ann. Phys.* (N.Y.), **25** 156
[44] Gómez Camacho A, Aguilera E F, Lubian J, Gomes P R S, 2013 *J. Phys.* G **40** 035103

Journal of Physics: Conference Series **492** (2014) 012006 doi:10.1088/1742-6596/492/1/012006

Quantum partner-dance in the $^{12}C + {}^{12}C$ system yields sub-Coulomb fusion resonances

Alexis Diaz-Torres[1], **Michael Wiescher**[2]

[1] ECT*, Strada delle Tabarelle 286, I-38123 Villazzano, Trento, Italy
[2] JINA and Department of Physics, University of Notre Dame, IN 46656, USA

E-mail: `torres@ectstar.eu`

Abstract. A preliminary study of the $^{12}C + {}^{12}C$ sub-Coulomb fusion reaction using the time-dependent wave-packet method is presented. The theoretical sub-Coulomb fusion resonances seem to correspond well with observations. The present method might be a more suitable tool for expanding the cross-section predictions towards lower energies than the commonly used potential-model approximation.

1. Introduction

The $^{12}C + {}^{12}C$ fusion reaction plays a key role in the chemical evolution of the Universe [1].The fusion cross section at energies near the Gamow peak ($\sim$ 1.5 MeV) is critical for the stellar carbon-burning rates, which is usually obtained by extrapolating high-energy fusion data [2, 3, 4], as direct measurements are very difficult to carry out at very low energies ($\leq$ 3 MeV). The observed fusion resonances make the extrapolation very uncertain [5, 6, 7].

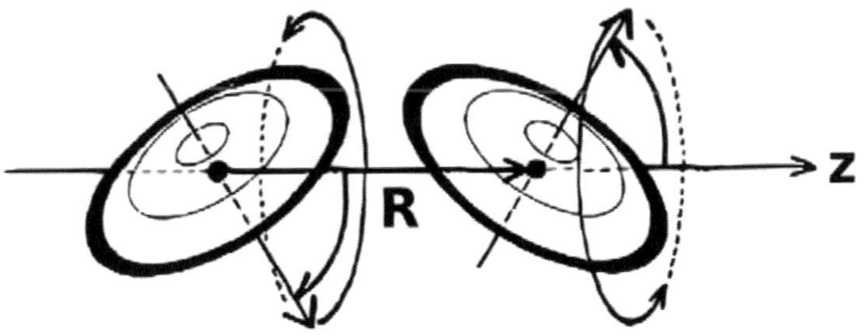

Figure 1. The $^{12}C + {}^{12}C$ nuclear molecule, exhibiting collective excitation modes.

The resonances may be mainly related to collective excitation modes in the dinuclear system, when the two ^{12}C nuclei come into contact (Fig. 1). The ^{12}C intrinsic symmetry axis vibrates and rotates with respect to the internuclear axis. The single-particle molecular shell structure at small internuclear distances is very sensitive to the alignment of the ^{12}C nuclei [8]. Non-axial symmetric configurations preserve the individuality of the overlapping nuclei, while this is not the case for the axial symmetric configuration. The former favors re-separation, and the latter fusion. The competition among these configurations, as a function of the incident energy and orbital angular momentum, should result in molecular resonance structures in the fusion excitation curve.

2. Method

This picture is here quantified using the time-dependent wave-packet (TDWP) method which has not been much exploited in nuclear physics [9] unlike chemical physics [10]. This method involves three steps [10]:

(1) the definition of the initial wave function $\Psi(t=0)$,

(2) the propagation $\Psi(0) \rightarrow \Psi(t)$, dictated by the time evolution operator, $\exp(-i\hat{H}t/\hbar)$, where $\hat{H}$ is the total Hamiltonian,

(3) the calculation of molecular collective states and the fusion cross section from the time-dependent wave function, $\Psi(t)$.

The wave function and the Hamiltonian are represented in a multi-dimensional numerical grid. In this work, these are considered a function of five collective coordinates (Fig. 1): the internuclear distance R, and the (θ_1, ϕ_1) and (θ_2, ϕ_2) spherical angles of the ^{12}C nuclei symmetry axis, thus reducing the complexity of the quantum many-body reaction problem. Moreover, the wave function is *not* expanded in any intrinsic basis (e.g., rotational or vibrational states of the individual nuclei), but it is calculated directly. The outgoing-wave-boundary condition at large internuclear distances as well as the *irreversible* process of fusion at small internuclear distances (usually described with an ingoing-wave-boundary condition) are here treated with the absorbing-boundary-condition [11]. The low-energy collision is described in the rotating center-of-mass frame within the nuclear molecular picture [12].

A simple case is the ^{16}O + ^{16}O collision, as the nuclei are spherical and inelastic coupled-channels effects are rather small, e.g, see Fig. 1 in Ref. [13]. It is a one-dimensional problem that involves only the internuclear-distance coordinate, R, whose solution within the present method can be compared to that of the time-independent Schrödinger equation with the usual boundary conditions [14, 15].

Figure 2 shows the transmission-coefficient excitation function for the total angular momentum $J = 0\hbar$, which is determined by two methods: solving the time-independent Schrödinger equation [15] (solid line) and employing the present method (squares). The agreement between the two methods is very good, demonstrating both the reliability of the time-dependent wave-packet method and how a single wave-packet propagation can provide accurate transmission coefficients over a range of collision energies [10].

3. Results and Discussion

Figure 3 shows cuts of the collective potential energy surface (PES) which is calculated using the finite-range liquid drop model [16] with nuclear shapes of the two-center shell model [8]. Fusion is here determined by two processes treated simultaneously: (i) the tunneling through the many Coulomb barriers in Fig. 3, and (ii) the ^{12}C nuclei re-alignment in the potential pockets. The 90-90 alignment (dashed line) dominates the Coulomb barriers penetrability, but the 0-0 alignment (solid line) is critical for fusion. Non-zero J partial waves substantially contribute to the fusion cross section, as the centrifugal contribution to the 90-90 Coulomb barrier is small for low $J \leq 6\hbar$ and the J-dependent Coriolis interaction, whose strength increases with J and the system compactness, strongly drives the non-axial symmetric configurations towards the potential pocket of the axial symmetric one (solid line), where fusion occurs.

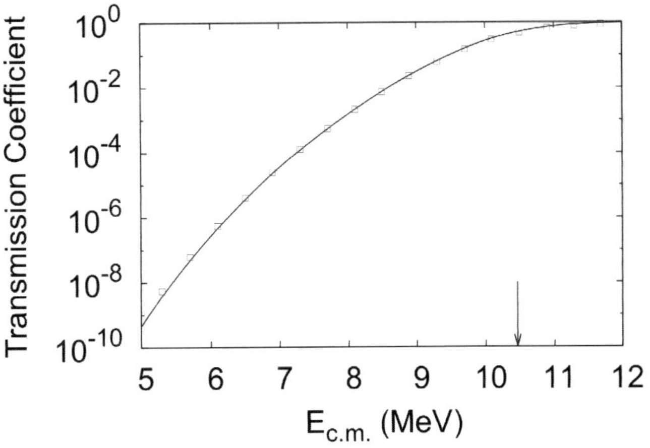

Figure 2. Excitation function of the transmission coefficient through the Coulomb barrier (the Broglia-Winther potential in Ref. [15]) for the ^{16}O + ^{16}O central collision, calculated with two methods: (1) solving the time-independent Schrödinger equation (solid line), and (2) the time-dependent wave-packet method (squares). The agreement between the two methods is very good. The arrow indicates the Coulomb-barrier energy.

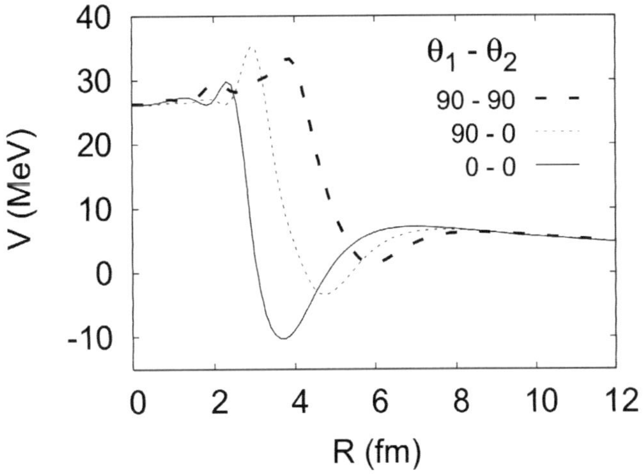

Figure 3. Cuts of the ^{12}C + ^{12}C collective PES, $V(R, \theta_1, \phi_1 = 0, \theta_2, \phi_2 = 0)$, as a function of the nuclei separation and alignment. The 90-90 alignment (dashed line) facilitates the access by tunneling to the potential pockets $(3 - 6$ fm). These are explored by the system, guided by the kinetic-energy operator [17]. The Coriolis force drives non-axial symmetric configurations towards the potential pocket of the axial symmetric configuration (solid line), where fusion occurs by a strong absorption.

The molecular resonance states with a total width of $\sim 100 - 180$ keV are presented in Fig. 4, which represent doorway states that irreversibly decay into more complex compound-nucleus states (simulated by the strong absorption providing the *inclusive* fusion probability) and the

$^{12}C + ^{12}C$ continuum scattering states.

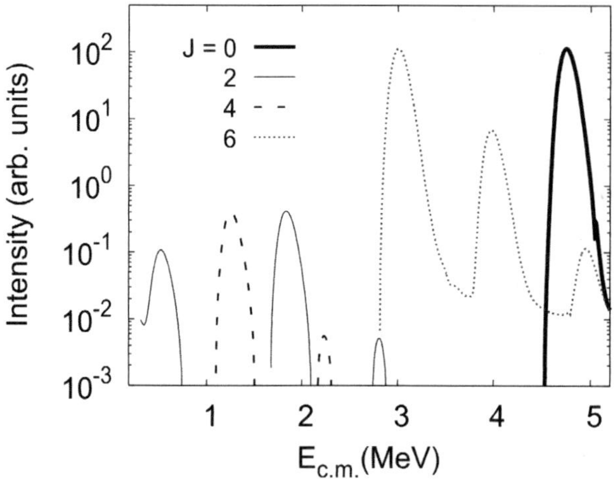

Figure 4. Energy spectrum of the $^{12}C + ^{12}C$ system in the potential pockets of Fig. 3. The peaks are molecular resonances with a given spin and total width of $\sim 100-180$ keV. The widths relate to the total time interval spent by the dinuclear system in the potential pockets. These resonances are doorway states for fusion, which decay into more complex compound-nucleus states and the $^{12}C + ^{12}C$ scattering states. Various resonances exist near the Gamow peak energy (1.5 MeV).

These molecular collective states are reflected in the total fusion excitation curve, as presented in Fig. 5 through the S-factor that includes $J \leq 8\hbar$. The fusion cross section, $\sigma_{fus}(E)$, is calculated taken into account the identity of the interacting nuclei and the parity of the radial wave function (only even partial waves J are included), i.e., $\sigma_{fus}(E) = \pi\hbar^2/(2\mu E)\sum_J(2J+1)(1+\delta_{1,2})P_J(E)$, where μ is the reduced mass, E is the incident energy in the center-of-mass frame and P_J is the partial fusion probability. The S-factor is $S(E) = \sigma_{fus}(E)E\exp(2\pi\eta)$, where the Sommerfeld parameter $\eta = (\mu/2)^{1/2}Z_1Z_2e^2/(\hbar E^{1/2})$ and Z_i is the charge number of the ^{12}C nuclei. The features resulting from the preliminary TDWP calculations (solid line) are consistent with those observed in the experimental data [2, 4] (squares). While the widths of the resonances show good agreement, the amplitudes are noticeably different: the theoretical predictions are too low in the higher energy range and too high in the lower energy range. This issue could be clarified using a bigger numerical grid (for checking convergence) and incorporating shell and pairing corrections into the collective mass and PES [15].

4. Summary

Using time-dependent wave-packet dynamics within a nuclear molecular picture, a quatitative study of the $^{12}C + ^{12}C$ sub-Coulomb fusion has been presented. Preliminary calculations are very promising, indicating a close correlation between molecular collective states and fusion. The present method might be a more suitable tool for expanding the cross section predictions towards lower energies than the usually employed potential-model approach.

Acknowledgments

Support through the Joint Institute for Nuclear Astrophysics JINA NSF Grant Phys-0822648 is acknowledged.

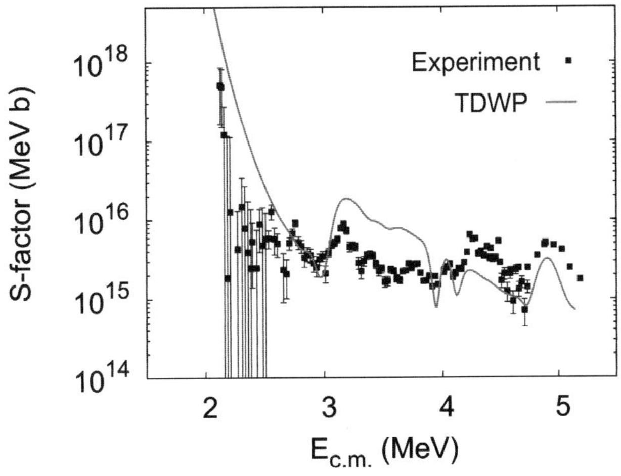

Figure 5. S-factor excitation function. There is a good *qualitative* agreement between measurements [2, 4] and preliminary TDWP calculations, highlighting the symphysis of molecular structure and fusion.

References

[1] Rolfs C and Rodney W S 1988 *Cauldrons in the Cosmos* (Chicago: University of Chicago Press)
[2] Aguilera E F *et al* 2006 *Phys. Rev. C* **73** 064601
[3] Barron-Palos L *et al* 2006 *Nucl. Phys. A* **779** 318
[4] Spillane T *et al* 2007 *Phys. Rev. Lett.* **98** 122501
[5] Gasques L R *et al* 2007 *Phys. Rev. C* **76** 035802
[6] Cooper R L *et al* 2009 *Ap. J* **702** 660
[7] Bennett M E *et al* 2010 *J. of Phys: Conf. Ser.* **202** 012023
[8] Diaz-Torres A 2008 *Phys. Rev. Lett.* **101** 122501
[9] Schneider J and Wolter H H 1991 *Z. Phys. A* **339** 177
[10] Tannor D J 2007 *Introduction to Quantum Mechanics: A Time Dependent Perspective* (Saulito: University Science Books)
[11] Ueda M, Yabana K and Nakatsukasa T 2004 *Nucl. Phys. A* **738** 288
[12] Greiner W, Park J Y and Scheid 1995 W *Nuclear Molecules* (Singapore: World Scientific Pub)
[13] Esbensen H 2008 *Phys. Rev. C* **77** 054608
[14] Hagino K, Rowley N and Kruppa A T 1999 *Comp. Phys. Comm.* **123** 143
[15] Diaz-Torres A, Gasques L R and Wiescher M 2007 *Phys. Lett. B* **652** 25
[16] Davies K T R and Nix J R 1976 *Phys. Rev. C* **14** 1977
[17] Gatti F et al 2005 *J. Chem. Phys.* **123** 174311

A solution to the cosmological ^{7}Li problem?

Osvaldo Civitarese and Mercedes Mosquera

Department of Physics, University of La Plata, Argentina

E-mail: osvaldo.civitarese@fisica.unlp.edu.ar; mmosquera@fcaglp.unlp.edu.ar

Abstract. In this work we report on the cosmological ^{7}Li problem, from a nuclear structure point of view, that is by including resonances in the calculation of the reactions which populate beryllium. It is found that the primordial abundance of lithium is reduced, as a consequence of the presence of resonant channels in the relevant cross sections. We establish constraints on the resonant energies, and make a comparison with the available observational data.

1. Introduction

BBN (for Big Bang Nucleosynthesis) is the framework which describes the formation of light nuclei in the early Universe. The theory relies upon nuclear reactions between primordial elements. It has only one parameter, the baryon density, which is related to the baryon-to-photon ratio. This parameter can be determined by the analysis of the data produced by the Wilkinson Microwave Anisotropy Probe (WMAP) satellite (1). The results are consistent with the observed abundances of deuterium and ^{4}He, but they disagree with the observed abundance of ^{7}Li. In the literature, this problem was analyzed using different theoretical approaches based on astrophysical considerations (2; 3; 4). Recently, the nuclear aspect of the problem concerning the abundance of ^{7}Li have been revisited (5; 6; 7), by considering nuclear-reaction and nuclear-structure mechanisms which may reduce the abundance of lithium due to a depletion of the production of beryllium.

In this work, we analyze the effect upon the production of ^{7}Li due to the inclusion of isolated resonances in the reactions involving ^{7}Be. To perform the calculations of the primordial abundances we use the formalism developed by (8), and to include the resonances we assume that the cross sections are described by the Breit-Wigner formula. We analyze the effects of the inclusion of these resonances separately, and, in particular, on the freeze-out temperature of beryllium, the ^{7}Be abundance and the lithium primordial abundance. We set constrains on the value of the resonant energy for each process using the available observable data.

2. Formalism

The nuclear reactions which are relevant for the present study are:

$$\begin{aligned} {}^{7}\mathrm{Be} + d &\to {}^{4}\mathrm{He} + {}^{4}\mathrm{He} + p \\ {}^{7}\mathrm{Be} + {}^{4}\mathrm{He} &\to \gamma + {}^{11}\mathrm{C} \\ {}^{7}\mathrm{Be} + n &\to p + {}^{7}\mathrm{Li}. \end{aligned} \tag{1}$$

The reaction rate $[ij;kl]$ for the process $i+j \rightarrow k+l$ [1]. is written (8)

$$[ij;kl] \quad = \quad \rho_B N_A \langle \sigma v \rangle = 0.93 \times 10^{-3} \Omega_B h^2 T_9^3 N_A \langle \sigma v \rangle \ , \tag{2}$$

where σ is the cross section, v is the relative velocity, $\rho_B = 0.93 \times 10^{-3} \Omega_B h^2 T_9^3$ is the density of baryonic matter, N_A is the Avogadro number per gram, T_9 is the temperature in units of $10^9 K$. The use of Maxwell-Boltzmann velocity-distribution leads to the estimate

$$\langle \sigma v \rangle \quad = \quad \left(\frac{\mu}{2\pi kT} \right)^{3/2} \int e^{-\frac{\mu v^2}{2kT}} v \sigma(E) d^3 v \ , \tag{3}$$

where μ is the reduced mass.

In order to take into account the resonances in the reaction rates, we assume that the cross section for a generic process is describe by the Breit-Wigner formula (9)

$$\sigma(E) \quad = \quad \frac{\pi \hbar^2}{2\mu E} \frac{\omega_r \Gamma_1 \Gamma_2}{(E - E_r)^2 + \Gamma^2/4} \ , \tag{4}$$

where Γ_i is the partial width for the decay of the resonant state, Γ is the sum over all partial widths, $\omega_r = \frac{(1+\delta_{ab}) g_r}{g_a g_b}$ and $g_r = 2J_r + 1$, J_r being the spin of the resonant state and E_r is the resonance energy in the center of momentum system. The average cross section $\langle \sigma v \rangle$ is

$$\langle \sigma v \rangle \quad = \quad \left(\frac{2\pi \hbar^2}{\mu kT} \right)^{3/2} \frac{(\omega \gamma)_r}{\hbar} e^{E_r/kT} \ , \tag{5}$$

where $\gamma_r = \left(\frac{\Gamma_1 \Gamma_2}{\Gamma} \right)_r$ The reaction rate is written

$$[ij;kl] = 9.69 \times 10^{-33} \Omega_B h^2 \left(\frac{T_9}{\mu} \right)^{3/2} (\omega \gamma)_r e^{-11.605 E_r / T_9} \ , \tag{6}$$

where T_9 is the temperature in units of 10^9 K. Empirical values of the widths may be obtained from nuclear reaction data. This is a sensitive aspect of the calculations, because the actual value of the resonances and their widths, extracted from experimental data, may restrict the range of values where the effect upon the abundance of ^{7}Li become noticeable (10). Having these limitations into account we shall explore the mechanism (that is the excitation and decay of resonances) for which the population of states leading to ^{7}Li may be depleted.

2.1. Calculation of primordial abundances

The equation that governs the primordial abundance Y_i of a given element i, is

$$\dot{Y}_i \quad = \quad J(t) - \Gamma(t) Y_i \tag{7}$$

The static solution of this equation is the quasi-static equilibrium (QSE) solution (8)

$$f_i \quad = \quad \frac{J(t)}{\Gamma(t)} . \tag{8}$$

In order to compute the primordial abundances, one must solve the system of differential equations, considering only the dominant reaction rates (8). The semi-analytical approach

[1] Hereafter, we denote the reactions rates as [Be d; $\alpha\, \alpha\, p$], [Be α; γ C] and [Be n; p 7], respectively, where p refers to proton, d to deuterium, α to ^{4}He, Be to ^{7}Be, γ to the photon, C to ^{11}C, n is the neutron and 7 stands for ^{7}Li.

consists in calculating the abundances between fixed points or stages. One solves the equations only for one element at each step and, for the other elements, one considers the quasi static equilibrium solution. The freeze-out of the production of each element takes places when the dominant destruction reaction rate equals the expansion rate of the Universe (11).

The differential equation that governs the primordial abundance of beryllium is

$$\begin{aligned} \dot{Y}_{\mathrm{Be}} &= Y_p Y_6[6\,p;\,\gamma\,\mathrm{Be}] + Y_3 Y_\alpha[3\,\alpha;\,\gamma\,\mathrm{Be}] - Y_\gamma Y_{\mathrm{Be}}[\mathrm{Be}\,\gamma;\,3\,\alpha] \\ &\quad - Y_n Y_{\mathrm{Be}}[\mathrm{Be}\,n;\,p\,7] - Y_p Y_{\mathrm{Be}}[\mathrm{Be}\,p;\,\gamma\,8] - Y_d Y_{\mathrm{Be}}[\mathrm{Be}\,d;\,\alpha\,\alpha\,p] \\ &\quad - Y_\alpha Y_{\mathrm{Be}}[\mathrm{Be}\,\alpha;\,\gamma\,\mathrm{C}]. \end{aligned} \tag{9}$$

According to (8), the predominant destruction reaction rate is $[\mathrm{Be}\,n;\,p\,7]$ and the predominant source term is $Y_3 Y_\alpha[3\,\alpha;\,\gamma\,\mathrm{Be}]$. The freeze-out temperature of ^{7}Be $\left(T_9^{\mathrm{Be}}\right)$ is calculated by equating the expansion rate of the Universe $\left(H = \frac{1}{356} T_9^2\right)$ with the dominant reaction (destruction) rate

$$H = Y_n[\mathrm{Be}\,n;\,p\,7] + f\left(E_r, T_9\right)\ , \tag{10}$$

where $f\left(E_r, T_9\right)$ stands for $Y_d[\mathrm{Be}\,d;\,\alpha\,\alpha\,p]$, and $Y_\alpha[\mathrm{Be}\,\alpha;\,\gamma\,\mathrm{C}]$. The neutron abundance can be determined by solving the quasi-static equation $\dot{Y}_n = 0$

$$Y_n = Y_d \frac{Y_d[d\,d;\,n\,3] + Y_t[d\,t;\,n\,\alpha]}{Y_p[n\,p;\,\gamma\,d] + Y_3[3\,n;\,p\,t] + [n]}\ . \tag{11}$$

In the last equation, t stands for tritium. Considering $\dot{Y}_t = 0$

$$Y_n = Y_d^2 \frac{[d\,d;\,n\,3] + [d\,d;\,p\,t]}{Y_p[n\,p;\,\gamma\,d] + Y_3[3\,n;\,p\,t] + [n]}\ . \tag{12}$$

Then, the quasi-static equation for ^{7}Be, we can compute the beryllium primordial abundance at the freeze-out temperature (8)

$$Y_{\mathrm{Be}} = \frac{Y_3 Y_\alpha[3\,\alpha;\,\gamma\,\mathrm{Be}]}{Y_n[\mathrm{Be}\,n;\,p\,7] + f\left(E_r, T_9^{\mathrm{Be}}\right)}\ . \tag{13}$$

The differential equation which governs the primordial abundance of lithium is

$$\dot{Y}_7 \;=\; Y_n Y_6[6\,n;\,\gamma\,7] + Y_n Y_{\mathrm{Be}}[\mathrm{Be}\,n;\,p\,7] + Y_t Y_\alpha[t\,\alpha;\,\gamma\,7] - Y_p Y_7[7\,p;\,\alpha\,\alpha]\ .$$

The quasi-static equation $\dot{Y}_7 = 0$ is solved by considering the dominant production and destruction terms

$$Y_7 = \frac{Y_n Y_{\mathrm{Be}}[\mathrm{Be}\,n;\,p\,7] + Y_t Y_\alpha[t\,\alpha;\,\gamma\,7]}{Y_p[t\,p;\,\alpha\,\alpha]}\ , \tag{14}$$

where Y_{Be} is the value obtained by evaluating Eq.(13) at the freeze-out temperature of beryllium. The abundance of tritium is determined by solving the equation $\dot{Y}_t = 0$, and the expression for Y_n is given in Eq. (12). In order to calculate the final lithium abundance, we evaluate this expression at the freeze-out temperature of lithium $\left(T_9^{\mathrm{Li}}\right)$ which is the solution of

$$Y_p[7\,p;\,\alpha\,\alpha] = H\ . \tag{15}$$

As one can see, the inclusion of an isolated resonance in either $[\mathrm{Be}\,d;\,\alpha\,\alpha\,p]$, $[\mathrm{Be}\,\alpha;\,\gamma\,\mathrm{C}]$ or $[\mathrm{Be}\,n;\,p\,7]$ does not affect the freeze-out temperature of lithium. However, its abundance is reduced due to the change in the beryllium primordial abundance resulting from the change in the freeze-out temperature of it.

3. Results

To compute the light nuclei abundances we have adopted the WMAP value of the baryon density $(\Omega_B h^2)_{\rm WMAP} = 0.0224 \pm 0.0008$ (1; 12; 13), the primordial abundances of proton, deuterium, ^{3}He and ^{4}He are

$$\begin{aligned} Y_p &= 0.749 \ , \\ Y_d &= 2.36 \times 10^{-5} \ , \\ Y_3 &= 6.76 \times 10^{-6} \ , \\ Y_\alpha &= 0.24915 \ , \end{aligned} \tag{16}$$

respectively. Thus, the freeze-out temperature of lithium (see Eq. (15)) is estimated at the value

$$T_9^{\rm Li} = 0.1849 \ . \tag{17}$$

As an example, we show in Figure 1 the normalized freeze-out temperature of beryllium $\left(\left(T_9^{\rm Be}\right)_{res}/T_9^{\rm Be}\right)$ as a function of the resonant energy, for different values of the parameter a. The freeze-out temperatures have been calculated with, $\left(T_9^{\rm Be}\right)_{res}$, and without, $T_9^{\rm Be}$, including a resonance in the spectrum of ^{7}Be. The temperature is lower than the standard one for resonant energies lower than 0.25 MeV, for all the values of the parameter a. The effect of the variation of a is noticeable. The freeze-out temperature decreases if a increases. It means that a resonant state with a large width, in the entrance channel of the reaction, will cause a significant reduction of the freeze-out temperature and the effect will be dominant at lower energies. Similarly, the

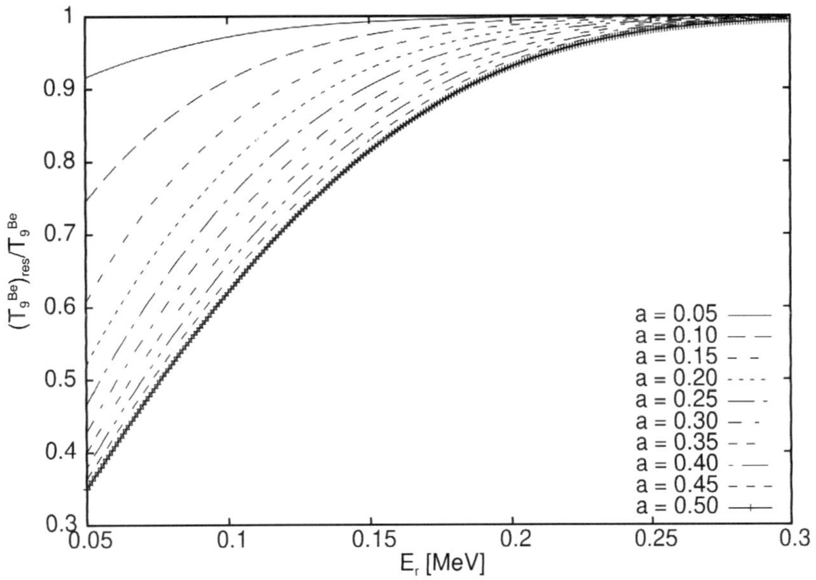

Figure 1. Ratio $\left(T_9^{\rm Be}\right)_{res}/T_9^{\rm Be}$ of the freeze-out temperature of beryllium as a function of the resonant energy E_r, for different values of the parameter a, considering an isolated resonance in the reaction $^7{\rm Be} + d \rightarrow ^4{\rm He} + ^4{\rm He} + p$

results for the primordial abundance of lithium with, $(Y_7)_{res}$, and without, Y_7, including a resonance in the spectrum of ^{7}Be are shown in Figure 2, as a function of the resonant energy, and for different values of the parameter a. The effect is quite noticeable. It is seen that the value of Y_7 can be reduced almost to one half of the standard value depending on the values of E_r and a. These results are in agreement with the results obtained by (5) and by (7).

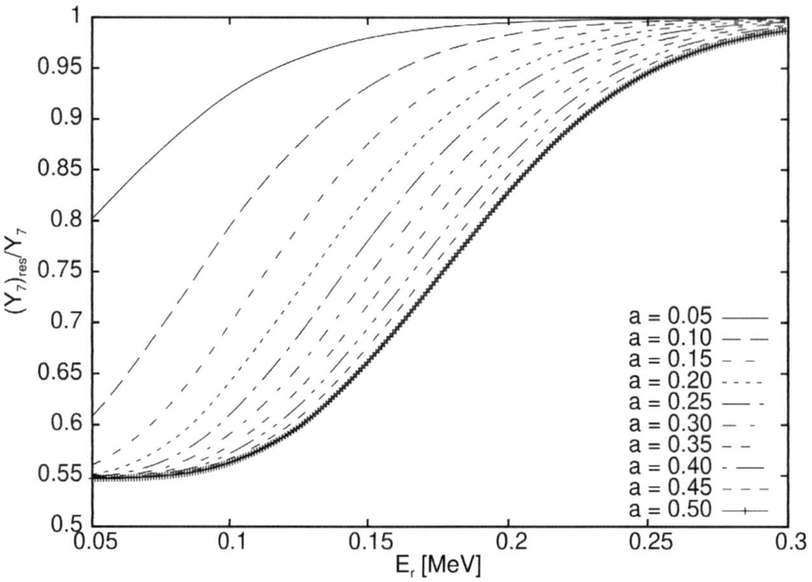

Figure 2. Ratio between the abundance of lithium with $((Y_7)_{res})$ and without including a resonant state (Y_7) in the beryllium spectrum, as a function of the resonant energy E_r, for different values of the parameter a, in the reaction $^7\mathrm{Be} + d \rightarrow {}^4\mathrm{He} + {}^4\mathrm{He} + p$

The available data of lithium may be used to to obtain the best-value of the resonant energy for different values of the parameter a, by implementing a χ^2-test. The results of the analysis are presented in Table 1. There is a good fit for all values of the parameter a considered,

Table 1. Best-fit and 1σ errors of the resonant energy of an isolated resonance in reaction $^7\mathrm{Be} + d \rightarrow {}^4\mathrm{He} + {}^4\mathrm{He} + p$, for different values of the parameter a (the parameter remains fixed in the statistical test).

a	$E_r \pm \sigma$ [MeV]	$\chi^2/(N-1)$
0.15	$0.066^{+0.014}_{-0.040}$	1.00
0.20	$0.080^{+0.016}_{-0.040}$	1.00
0.25	$0.091^{+0.016}_{-0.040}$	1.00
0.30	$0.099^{+0.017}_{-0.043}$	1.00
0.35	$0.105^{+0.018}_{-0.044}$	1.00
0.40	$0.111^{+0.018}_{-0.046}$	1.00
0.45	$0.116^{+0.018}_{-0.048}$	1.00
0.50	$0.120^{+0.019}_{-0.049}$	1.00

since $\chi^2/(N-1) = 1$ (N is the number of observational data, $N = 11$). The best-fit value of the resonant energy increases its value for larger values of the parameter a. These results for the resonant energy are dependent on the observable data set used in the statistical analysis, however, the favored value of the resonant energy is lower than 0.2 MeV.

4. Conclusion

In this work we have computed the primordial abundances of light elements using a semi-analytical approach and considering the inclusion of isolated resonances in the reactions leading to beryllium. We have found that there exists a reduction of the primordial abundance of lithium due to a depletion of the beryllium abundance, produced by resonances in the reactions, particularly in $^7\mathrm{Be} + d \rightarrow ^4\mathrm{He} + ^4\mathrm{He} + p$ and $^7\mathrm{Be} + ^4\mathrm{He} \rightarrow \gamma + ^{11}\mathrm{C}$. However, if the resonance is present in the reaction $^7\mathrm{Be} + n \rightarrow p + ^7\mathrm{Li}$, the amount of primordial lithium is not modified substantially, a result which is also in agreement with the results of Refs. (5; 7). We have performed a statistical analysis in order to obtain the best-fit value of the resonant energy, and found that, for the reaction $^7\mathrm{Be} + d \rightarrow ^4\mathrm{He} + ^4\mathrm{He} + p$, the best-fit value of E_r lies between 50 to 150 keV (given a partial width in the range of 10 to 60 keV). The resonant energy for $^7\mathrm{Be} + ^4\mathrm{He} \rightarrow \gamma + ^{11}\mathrm{C}$ is higher, between 250 to 350 keV (given partial width in the range of 30 to 164 keV).

5. Acknowledgments

Support for this work was provided by the PIP 0740 of the National Research Council (CONICET) of Argentina, and by the ANPCYT of Argentina. The authors are members of the Scientific Research Career of the CONICET. Discussions with Prof. R. J. Liotta (KTH, Stockholm) are gratefully acknowledged.

References

[1] Larson D *et al.* 2011 *Astrophys. J. S.* **192** 16
[2] Richard O, Michaud G and Richer J 2005 *Astrophys. J.* **619** 538–548
[3] Meléndez J *et al.* 2010 *A & A* **515** L3
[4] Lind K *et al.* 2010 *IAU Symposium* (*IAU Symposium* vol 268) ed C Charbonnel, M Tosi, F Primas, & C Chiappini pp 263–268
[5] Broggini C, Canton L, Fiorentini G and Villante F L 2012 *JCAP* **6** 30
[6] Kirsebom O S and Davids B 2011 *Phys. Rev. C* **84** 058801
[7] Cyburt R H and Pospelov M 2012 *International Journal of Modern Physics E* **21** 50004
[8] Esmailzadeh R, Starknam G D and Dimopoulos S 1991 *Astrophys. J.* **378** 504–518
[9] Fowler W A, Caughlan G R and Zimmerman B A 1975 *Ann. Rev. Astron. Astroph* **13** 69
[10] O'Malley P D *et al.* 2011 *Physical Review C* **84** (R) 042801
[11] Bernstein J, Brown L S and Feinberg G 1989 *Reviews of Modern Physics* **61** 25–39
[12] Spergel D N *et al.* 2003 *Astrophys. J. Suppl. Ser.* **148** 175–194
[13] Spergel D N *et al.* 2007 *Astrophys. J. Suppl. Ser.* **170** 377–408

Journal of Physics: Conference Series **492** (2014) 012008 doi:10.1088/1742-6596/492/1/012008

Pygmies, giants, and skins as laboratory constraints on the equation of state of neutron-rich matter

J Piekarewicz
Department of Physics, Florida State University, Tallahassee, FL 32306-4350, USA
E-mail: jpiekarewicz@fsu.edu

Abstract. Laboratory experiments sensitive to the density dependence of the symmetry energy may place stringent constraints on the equation of state of neutron-rich matter and, thus, on the structure, dynamics, and composition of neutron stars. Understanding the equation of state of neutron-rich matter is a central goal of nuclear physics that cuts across a variety of disciplines. In this contribution I focus on how laboratory experiments on neutron skins and on both Pygmy and Giant resonances can help us elucidate the structure of neutron stars.

1. Introduction

One of the central questions framing the recent report by The Committee on the Assessment of and Outlook for Nuclear Physics is *"How does subatomic matter organize itself?"* [1]. Remarkably, most of the fascinating phases that are predicted to emerge in this subatomic domain can not be probed under normal laboratory conditions. However, such novel states of matter become stable in the interior of neutron stars by virtue of the enormous gravitational fields. The fascinating phases predicted to exist in the crust of neutron stars, such as Coulomb crystals of neutron-rich nuclei and nuclear pasta, are within the purview of the Facility for Rare Isotope Beams (FRIB) which has as one of its science goals to provide an understanding of matter in the crust of neutron stars [2]. FRIB is of relevance to neutron-star structure because at sub-saturation densities the uniform neutron-rich matter residing in the stellar core becomes unstable against cluster formation. That is, at these sub-saturation densities the separation between nucleons increases to such an extent that it becomes energetically favorable for the system to segregate into regions of normal density (*i.e.*, nuclear clusters) embedded in a dilute, likely superfluid, neutron vapor. This *"clustering instability"* signals the transition from the uniform liquid core to the non-uniform stellar crust (see Fig. 1).

The outer stellar crust of relevance to the FRIB program is comprised of a Coulomb lattice of neutron-rich nuclei immersed in a uniform electron gas [3, 4, 5, 6]). At the lowest densities of the outer crust, the Coulomb lattice is formed from ^{56}Fe nuclei. However, as the density increases—and given that the electronic Fermi energy increases rapidly with density—it becomes energetically favorable for electrons to capture into protons leading to the formation of a Coulomb crystal of progressively more neutron-rich nuclei; a progression that starts with ^{56}Fe and is predicted to terminate with the exotic, neutron-rich nucleus ^{118}Kr (see Fig. 1).

Eventually, the neutron-proton asymmetry becomes too large for the nuclei to bind any more neutrons and the excess neutrons go into the formation of a dilute neutron vapor. However, at inner-crust densities, distance scales that were well separated in both the crystalline phase (where

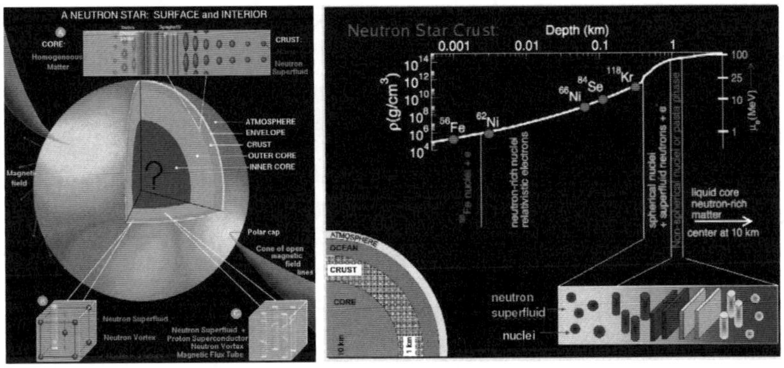

Figure 1. A scientifically-accurate rendition of the structure and phases of a neutron star (courtesy of Dany Page) and the composition of the stellar crust (courtesy of Sanjay Reddy).

the long-range Coulomb interaction dominates) and in the uniform phase (where the short-range strong interaction dominates) become comparable. This gives rise to *"Coulomb frustration"*, a phenomenon characterized by the formation of a myriad of complex structures radically different in topology yet extremely close in energy. Given that these complex structures—collectively referred to as *"nuclear pasta"*—are very close in energy, it has been speculated that the transition from the highly ordered crystal to the uniform phase must proceed through a series of changes in the dimensionality and topology of these structures [7, 8]. Moreover, due to the preponderance of low-energy states, frustrated systems display an interesting and unique low-energy dynamics that has been studied using a variety of techniques including numerical simulations [9, 10, 11, 12, 13, 14].

FRIB will not be the only experimental program of direct relevance to the structure of neutron stars. Indeed, the *Lead Radius EXperiment* ("PREX") at the Jefferson Laboratory measures the neutron radius of ^{208}Pb. When combining this purely electroweak result with the accurately known charge radius of ^{208}Pb, one obtains its *"neutron skin"*—the difference between the root-mean-square neutron and proton radii. As we shall see later, the neutron skin correlates strongly to the pressure of pure neutron matter at saturation density which, in turn, correlates strongly to the neutron-star radius.

2. Formalism

Neutron stars satisfy the Tolman-Oppenheimer-Volkoff (TOV) equations, which are the extension of Newton's laws to the domain of general relativity. The TOV equations may be expressed as a coupled set of first-order differential equations of the following form:

$$\frac{dP}{dr} = -G\frac{\mathcal{E}(r)M(r)}{r^2}\left[1+\frac{P(r)}{\mathcal{E}(r)}\right]\left[1+\frac{4\pi r^3 P(r)}{M(r)}\right]\left[1-\frac{2GM(r)}{r}\right]^{-1}, \tag{1}$$

$$\frac{dM}{dr} = 4\pi r^2\mathcal{E}(r)\,, \tag{2}$$

where G is Newton's gravitational constant and $P(r)$, $\mathcal{E}(r)$, and $M(r)$ represent the pressure, energy density, and enclosed-mass profiles of the star, respectively. Note that the three terms enclosed in square brackets in Eq. (1) are of general-relativistic origin. Notably, the only input that neutron stars are sensitive to is the equation of state (EOS), namely, $P = P(\mathcal{E})$. In Fig. 2 we display *mass-vs-radius* relations as predicted by three relativistic mean-field models [15]. Surprisingly, all three models—NL3 [16, 17], FSU [18], and IU-FSU [15], are able to accurately reproduce a variety of ground-state observables throughout the nuclear chart. Yet, the predictions displayed in Fig. 2 are significantly different. A central goal of this contribution is to identify critical laboratory observables that may be used to constrain the structure, dynamics, and composition of neutron stars.

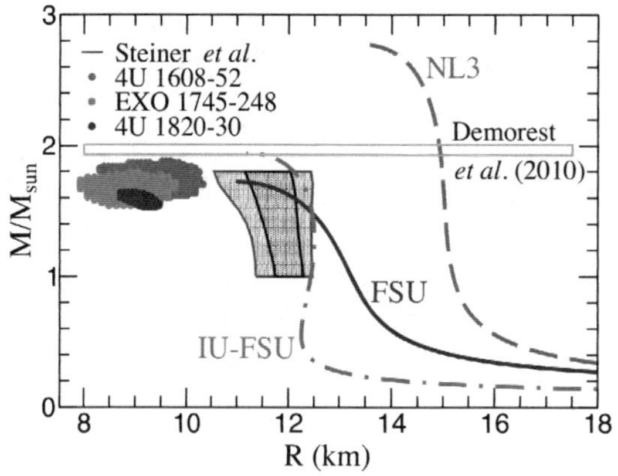

Figure 2. (Color online) *Mass-vs-Radius* relation predicted by the three relativistic mean-field models [15]. The observational data that suggest very small stellar radii represent 1σ confidence contours for the three neutron stars reported in Ref. [19]. The two shaded areas that suggest larger radii are 1σ and 2σ contours extracted from the analysis of Ref. [20]. The seminal measurement of the heaviest neutron star by Demorest et al., is also indicated in the figure [21].

The starting point for the calculation of both nuclear and neutron-star structure is a relativistic energy density functional characterized by the interacting Lagrangian density of Ref. [22] supplemented by an isoscalar-isovector term first introduced in Ref. [23]. That is,

$$\begin{aligned}\mathscr{L}_{\rm int} &= \bar{\psi}\left[g_{\rm s}\phi - \left(g_{\rm v}V_\mu + \frac{g_\rho}{2}\boldsymbol{\tau}\cdot\mathbf{b}_\mu + \frac{e}{2}(1+\tau_3)A_\mu\right)\gamma^\mu\right]\psi \\ &- \frac{\kappa}{3!}(g_{\rm s}\phi)^3 - \frac{\lambda}{4!}(g_{\rm s}\phi)^4 + \frac{\zeta}{4!}g_{\rm v}^4(V_\mu V^\mu)^2 + \Lambda_{\rm v}\left(g_\rho^2\,\mathbf{b}_\mu\cdot\mathbf{b}^\mu\right)\left(g_{\rm v}^2 V_\nu V^\nu\right). \qquad (3)\end{aligned}$$

The Lagrangian density includes an isodoublet nucleon field (ψ) interacting via the exchange of two isoscalar mesons, a scalar (ϕ) and a vector (V^μ), one isovector meson (b^μ), and the photon (A^μ) [24, 25]. In addition to meson-nucleon interactions, the Lagrangian density is supplemented by four nonlinear meson interactions with coupling constants (κ, λ, ζ, and $\Lambda_{\rm v}$) that are included primarily to soften the equation of state of both symmetric nuclear matter and pure neutron matter. For a detailed discussion on the impact of these terms on various quantities of theoretical, experimental, and observational interest see Ref. [26].

Laboratory experiments may play a critical role in constraining the size of neutron stars because stellar radii are controlled by the density dependence of the symmetry energy in the immediate vicinity of nuclear-matter saturation density [27]. Recall that the symmetry energy may be viewed as the difference in the energy between pure neutron matter and symmetric nuclear matter. A particularly critical property of the symmetry energy that is poorly constrained is its slope at saturation density—a quantity customarily denoted by L and closely related to the pressure of pure neutron matter [28]. Although L is not directly observable, it is strongly correlated to the thickness of the neutron skin of heavy nuclei [29, 30]. As indicated in the left-hand panel of Fig. 3, the thickness of the neutron skin depends sensitively on the pressure of neutron-rich matter: *the greater the pressure the thicker the neutron skin*. And it is exactly this same pressure that supports neutron stars against gravitational collapse. Thus, as indicated in the right-hand panel of Fig. 3, models with thicker neutron skins often produce neutron stars with larger radii [23, 31]. Thus, it is possible to study "data-to-data" relations between the neutron-rich skin of a heavy nucleus and the radius of a neutron star.

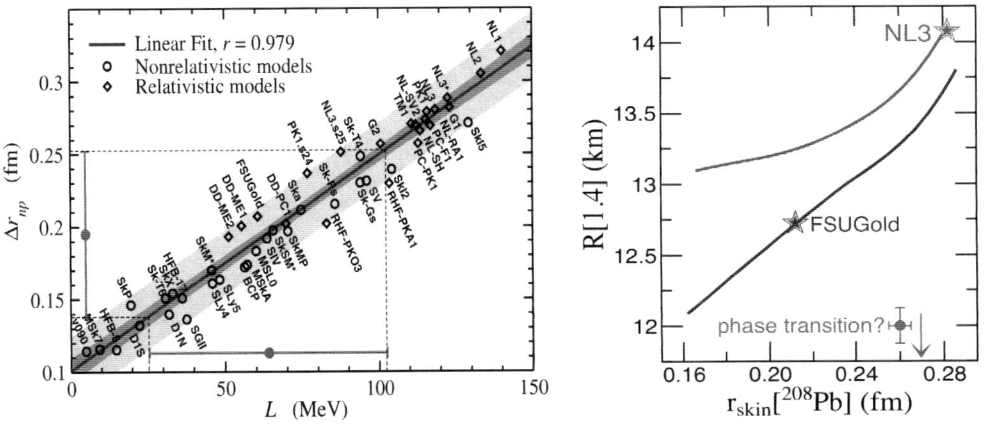

Figure 3. (Color online) The left-hand panel displays the correlation between the neutron-skin of ^{208}Pb and the slope of the symmetry energy for a variety of nonrelativistic and relativistic models [32]. The right-hand panel shows the correlation between the neutron-skin of ^{208}Pb and the radius of a $1.4\,M_\odot$ neutron star for two relativistic mean-field models.

3. Neutron Skins: The Lead Radius Experiment (PREX)

The successfully commissioned Lead Radius Experiment has provided the first model-independent evidence of the existence of a significant neutron skin in ^{208}Pb [33, 34]. Parity violation at low momentum transfers is particularly sensitive to the neutron distribution because the neutral weak-vector boson (Z^0) couples preferentially to the neutrons in the target [35]. Although PREX achieved the systematic control required to perform this challenging experiment, unforeseen technical problems resulted in time losses that significantly compromised the statistical accuracy of the measurement. This resulted in the following value for the neutron-skin thickness of ^{208}Pb [33, 34]:

$$r_{\rm skin} = R_n - R_p = 0.33^{+0.16}_{-0.18}\ {\rm fm}. \tag{4}$$

Given that the determination of the neutron radius of a heavy nucleus is a problem of fundamental importance with far reaching implications in areas as diverse as nuclear structure [29, 30, 36, 37, 38], atomic parity violation [39, 40], heavy-ion collisions [41, 42, 43, 44, 45], and neutron-star structure [23, 31, 46, 47, 48, 49, 50], the PREX collaboration has made a successful proposal for additional beam time so that the original 1% goal (or ±0.05 fm) may be attained [51]. While the scientific case for such a critical experiment remains strong, the search for additional physical observables that may be both readily accessible and strongly correlated to the neutron skin (and thus also to L) is a worthwhile enterprise. It is precisely the exploration of such a correlation between the *electric dipole polarizability* and the neutron-skin thickness of ^{208}Pb that is at the center of the next section.

4. Pygmies and Giant Resonances

A promising complementary approach to the parity-violating program relies on the electromagnetic excitation of the electric dipole mode [52]. For this mode of excitation—

perceived as a collective oscillation of neutrons against protons—the symmetry energy acts as the restoring force. In particular, models with a soft symmetry energy predict large values for the symmetry energy at the densities of relevance to the excitation of this mode. As a consequence, the stronger restoring force of the softer models generates a dipole response that is both hardened and quenched relative to its stiffer counterparts. In particular, the *inverse* energy-weighted sum, which is directly proportional to the electric dipole polarizability $\alpha_{\rm D}$, is highly sensitive to the density dependence of the symmetry energy [53]. This sensitivity suggests the existence of the following interesting correlation in heavy nuclei: *the larger* $r_{\rm skin}$ *the larger* α_D.

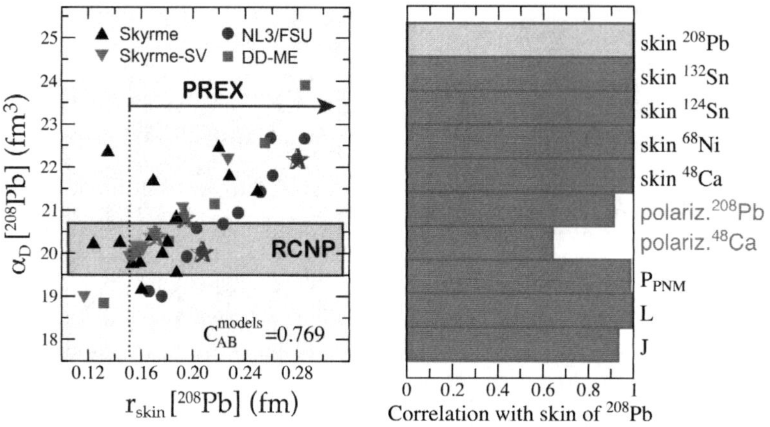

Figure 4. (Color online) Predictions from a variety of nuclear models for the electric dipole polarizability and neutron-skin thickness of ^{208}Pb are shown on the left-hand side of the figure. Also shown are constrains on the neutron-skin thickness from PREX [33, 34] and on the dipole polarizability from RCNP [55, 56]. On the right-hand side of the figure we show correlation coefficients between the neutron-skin thickness of ^{208}Pb and several observables as obtained from a covariance analysis based on the FSU interaction [59].

To test the validity of this correlation we display on the left-hand panel of Fig. 4 $\alpha_{\rm D}$ in ^{208}Pb as a function of its corresponding neutron-skin as predicted by a large number (48) of nuclear-structure models [54]. From the distribution of electric dipole strength ($R_{\rm E1}$) the dipole polarizability is readily extracted from the inverse energy-weighted sum. That is,

$$\alpha_{\rm D} = \frac{8\pi}{9}e^2 \int_0^\infty \omega^{-1} R_{\rm E1}(\omega)\, d\omega\,. \tag{5}$$

At first glance a clear (positive) correlation between the dipole polarizability and the neutron skin is discerned. However, on closer examination one observes a significant scatter in the results—especially in the case of the standard Skyrme forces (denoted by the black triangles). In particular, by including the predictions from all the 48 models under consideration, a correlation coefficient of 0.77 was obtained. Also shown in the figure are experimental constraints imposed from PREX and the recent high-resolution measurement of $\alpha_{\rm D}$ in ^{208}Pb [55, 56]. By imposing these recent experimental constraints, several of the models—especially those with either a very soft or very stiff symmetry energy—may already be ruled out.

However, to establish how the dipole polarizability may provide a unique constraint on the neutron-skin thickness of neutron-rich nuclei and other isovector observables we display on the right-hand panel of Fig. 4 correlation coefficients computed using a single underlying model, namely, FSU [18]. For details on the implementation of the required covariance analysis we refer the reader to Refs. [57, 58, 59]. According to the model, an accurate measurement of the neutron skin-thickness in ^{208}Pb significantly constrains the neutron skin on a variety of other neutron-rich nuclei. Moreover, the correlation coefficient between the neutron skin and $\alpha_{\rm D}$ in ^{208}Pb is very large (of about 0.9). This suggests that a multi-prong approach consisting of combined measurements of both neutron skins and $\alpha_{\rm D}$—ideally on a variety of nuclei—should significantly constrain the isovector sector of the nuclear energy density functional as well as the EOS of neutron-rich matter.

Naturally, a more stringent constrain on the isovector sector of the nuclear density functional is expected to emerge along an isotopic chain as the nucleus develops a neutron-rich skin. Concomitant with the development of a neutron skin one expects the emergence of low energy dipole strength—the so-called *pygmy dipole resonance* [60, 61, 62, 63, 64, 65, 66]. Thus, it has been suggested that the pygmy dipole resonance (PDR)—speculated to be an excitation of the neutron-rich skin against the isospin symmetric core—may be used as a constraint on the neutron skin [67].

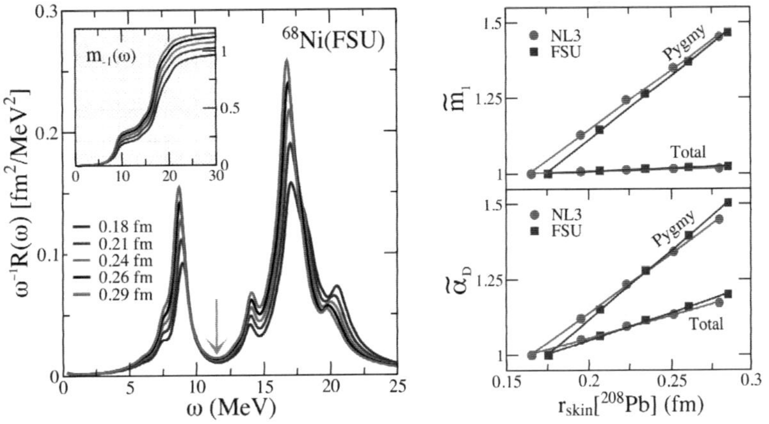

Figure 5. (Color online) The inverse energy weighted dipole response in ^{68}Ni computed with the FSU family of effective interactions is shown on the left-hand side of the figure. The inset displays the running sum. The arrow indicates the (ad-hoc) energy at which the low-energy (pygmy) response is separated from the high-energy (giant) response. On the right-hand side the fractional change in the energy weighted sum and dipole polarizability for ^{68}Ni are displayed as a function of the neutron-skin thickness of ^{208}Pb. See Ref. [53] for more details.

In particular, the fraction of the *inverse energy weighted sum rule* concentrated on the low-energy region appears to be sensitive to the neutron skin of neutron-rich nuclei. The inverse energy weighted response $\omega^{-1}R(\omega)$ is displayed on the left-hand panel of Fig. 5 for the neutron-rich nucleus ^{68}Ni. Given that the ω^{-1} factor enhances the low-energy part of response, the Pygmy resonance accounts for a significant fraction (of about 20-25%) of m_{-1} (which is proportional to $\alpha_{\rm D}$). Pictorially, this behavior is best illustrated in the inset of Fig. 5 which displays the *"running"* $m_{-1}(\omega)$ sum. The inset provides a clear indication that both the total m_{-1} moment as well as the fraction contained in the Pygmy resonance are highly sensitive to the neutron-skin thickness of ^{208}Pb. To heighten this sensitivity we display on the right-hand panel of Fig. 5 the *fractional change* in both the total and Pygmy contributions to the m_1 moment (*i.e.*, the energy weighted sum rule) and to the dipole polarizability α_D as a function of the neutron skin of ^{208}Pb (we denote these fractional changes with a *"tilde"* in the figure). These results illustrate the strong correlation between the neutron skin and α_D and establish how a combined measurement of these laboratory observables will be of vital importance in constraining the isovector sector of the nuclear density functional.

5. Conclusions

Measurements of neutron radii provide important constraints on the isovector sector of nuclear density functionals and offer vital guidance in areas as diverse as atomic parity violation, heavy-ion collisions, and neutron-star structure. In this contribution we examined the possibility of using the quintessential nuclear mode—the isovector dipole resonance—as a promising complementary observable. For this mode of excitation in which protons oscillate coherently against neutrons, the symmetry energy acts as its restoring force. Thus, models with a soft symmetry energy predict large values for the symmetry energy at the densities of relevance to the excitation of this mode. As a consequence, softer models generates a dipole response that is both hardened and quenched relative to the stiffer models. However, being protected by the Thomas-Reiche-Kuhn sum rule, the energy weighted sum rule is largely insensitive to this behavior. In contrast, for the inverse energy-weighted sum—which is directly proportional to the electric dipole polarizability $\alpha_{\rm D}$—the quenching and hardening act in tandem. Thus, models with a soft symmetry energy predict smaller values of $\alpha_{\rm D}$ than their stiffer counterparts. This results in a powerful "data-to-data" relation: *the smaller* $\alpha_{\rm D}$ *the thinner the neutron skin*. Moreover, we saw that a significant amount of electric dipole strength is concentrated in the low-energy fragment; the so-called Pygmy resonance.

In summary, motivated by two seminal experiments [33, 55], we examined possible correlations between the electric dipole polarizability and the neutron skin of neutron-rich nuclei. The neutron-skin thickness of a heavy nucleus is a quantity of critical importance for our understanding of a variety of nuclear and astrophysical phenomena. In particular, the neutron-skin thickness of ^{208}Pb can provide stringent constrains on the density dependence of the symmetry energy which, in turn, has a strong impact on the structure, dynamics, and composition of neutron stars. We conclude that precise measurements of neutron skins and $\alpha_{\rm D}$—ideally on a variety of nuclei— should significantly constrain the isovector sector of the nuclear energy density functional and will provide critical insights into the nature of neutron-rich matter.

Acknowledgments

This work was supported in part by grant DE-FD05-92ER40750 from the Department of Energy.

References

[1] The Committee on the Assessment of and Outlook for Nuclear Physics; Board on Physics and Astronomy; Division on Engineering and Physical Sciences; National Research Council, *Nuclear Physics: Exploring the Heart of Matter* (The National Academies Press, 2012), ISBN 9780309260404, URL `http://www.nap.edu/openbook.php?record_id=13438`.
[2] Nuclear Science Advisory Committee, *2007 Long Range Plan: The Frontiers of Nuclear Science*, URL `http://science.energy.gov/np/nsac`.
[3] G. Baym, C. Pethick, and P. Sutherland, Astrophys. J. **170**, 299 (1971).
[4] S. B. Ruester, M. Hempel, and J. Schaffner-Bielich, Phys. Rev. **C73**, 035804 (2006).
[5] X. Roca-Maza and J. Piekarewicz, Phys. Rev. **C78**, 025807 (2008).
[6] X. Roca-Maza, J. Piekarewicz, T. Garcia-Galvez, and M. Centelles (2011a), contribution to the book *Neutron Star Crust* (Nova Publishers), `1109.3011`.
[7] D. G. Ravenhall, C. J. Pethick, and J. R. Wilson, Phys. Rev. Lett. **50**, 2066 (1983).
[8] M. Hashimoto, H. Seki, and M. Yamada, Prog. Theor. Phys. **71**, 320 (1984).
[9] C. J. Horowitz, M. A. Perez-Garcia, and J. Piekarewicz, Phys. Rev. **C69**, 045804 (2004a).
[10] C. J. Horowitz, M. A. Perez-Garcia, J. Carriere, D. K. Berry, and J. Piekarewicz, Phys. Rev. **C70**, 065806 (2004b).
[11] C. J. Horowitz, M. A. Perez-Garcia, D. K. Berry, and J. Piekarewicz, Phys. Rev. **C72**, 035801 (2005).

[12] G. Watanabe, K. Sato, K. Yasuoka, and T. Ebisuzaki, Phys. Rev. **C68**, 035806 (2003).
[13] G. Watanabe, T. Maruyama, K. Sato, K. Yasuoka, and T. Ebisuzaki, Phys. Rev. Lett. **94**, 031101 (2005).
[14] G. Watanabe, H. Sonoda, T. Maruyama, K. Sato, K. Yasuoka, et al., Phys. Rev. Lett. **103**, 121101 (2009).
[15] F. J. Fattoyev, C. J. Horowitz, J. Piekarewicz, and G. Shen, Phys. Rev. **C82**, 055803 (2010).
[16] G. A. Lalazissis, J. Konig, and P. Ring, Phys. Rev. **C55**, 540 (1997).
[17] G. A. Lalazissis, S. Raman, and P. Ring, At. Data Nucl. Data Tables **71**, 1 (1999).
[18] B. G. Todd-Rutel and J. Piekarewicz, Phys. Rev. Lett **95**, 122501 (2005).
[19] F. Ozel, G. Baym, and T. Guver, Phys. Rev. **D82**, 101301 (2010).
[20] A. W. Steiner, J. M. Lattimer, and E. F. Brown, Astrophys. J. **722**, 33 (2010).
[21] P. Demorest, T. Pennucci, S. Ransom, M. Roberts, and J. Hessels, Nature **467**, 1081 (2010).
[22] H. Mueller and B. D. Serot, Nucl. Phys. **A606**, 508 (1996).
[23] C. J. Horowitz and J. Piekarewicz, Phys. Rev. Lett. **86**, 5647 (2001a).
[24] B. D. Serot and J. D. Walecka, Adv. Nucl. Phys. **16**, 1 (1986).
[25] B. D. Serot and J. D. Walecka, Int. J. Mod. Phys. **E6**, 515 (1997).
[26] J. Piekarewicz, Phys. Rev. **C76**, 064310 (2007).
[27] J. M. Lattimer and M. Prakash, Phys. Rept. **442**, 109 (2007).
[28] J. Piekarewicz and M. Centelles, Phys. Rev. **C79**, 054311 (2009).
[29] B. A. Brown, Phys. Rev. Lett. **85**, 5296 (2000).
[30] R. J. Furnstahl, Nucl. Phys. **A706**, 85 (2002).
[31] C. J. Horowitz and J. Piekarewicz, Phys. Rev. **C64**, 062802 (2001b).
[32] X. Roca-Maza, M. Centelles, X. Viñas, and M. Warda, Phys. Rev. Lett. **106**, 252501 (2011b).
[33] S. Abrahamyan, Z. Ahmed, H. Albataineh, K. Aniol, D. Armstrong, et al., Phys. Rev. Lett. **108**, 112502 (2012).
[34] C. Horowitz, Z. Ahmed, C. Jen, A. Rakhman, P. Souder, et al., Phys. Rev. **C85**, 032501 (2012).
[35] T. Donnelly, J. Dubach, and I. Sick, Nucl. Phys. **A503**, 589 (1989).
[36] P. Danielewicz, Nucl. Phys. **A727**, 233 (2003).
[37] M. Centelles, X. Roca-Maza, X. Viñas, and M. Warda, Phys. Rev. Lett. **102**, 122502 (2009).
[38] M. Centelles, X. Roca-Maza, X. Viñas, and M. Warda, Phys. Rev. **C82**, 054314 (2010).
[39] S. J. Pollock, E. N. Fortson, and L. Wilets, Phys. Rev. **C46**, 2587 (1992).
[40] T. Sil, M. Centelles, X. Viñas, and J. Piekarewicz, Phys. Rev. **C71**, 045502 (2005).
[41] M. B. Tsang et al., Phys. Rev. Lett. **92**, 062701 (2004).
[42] L.-W. Chen, C. M. Ko, and B.-A. Li, Phys. Rev. Lett. **94**, 032701 (2005).
[43] A. W. Steiner and B.-A. Li, Phys. Rev. **C72**, 041601 (2005).
[44] D. V. Shetty, S. J. Yennello, and G. A. Souliotis, Phys. Rev. **C76**, 024606 (2007).
[45] M. B. Tsang et al., Phys. Rev. Lett. **102**, 122701 (2009).
[46] C. J. Horowitz and J. Piekarewicz, Phys. Rev. **C66**, 055803 (2002).
[47] J. Carriere, C. J. Horowitz, and J. Piekarewicz, Astrophys. J. **593**, 463 (2003).
[48] A. W. Steiner, M. Prakash, J. M. Lattimer, and P. J. Ellis, Phys. Rept. **411**, 325 (2005).
[49] B.-A. Li and A. W. Steiner, Phys. Lett. **B642**, 436 (2006).
[50] F. J. Fattoyev and J. Piekarewicz, Phys. Rev. **C82**, 025810 (2010).
[51] K. Paschke, K. Kumar, R. Michaels, P. A. Souder, and G. M. Urciuoli (2012), URL `http://hallaweb.jlab.org/parity/prex/prexII.pdf`.
[52] M. N. Harakeh and A. van der Woude, *Giant Resonances-Fundamental High-frequency Modes of Nuclear Excitation* (Clarendon, Oxford, 2001).
[53] J. Piekarewicz, Phys. Rev. **C83**, 034319 (2011).
[54] J. Piekarewicz, B. Agrawal, G. Colò, W. Nazarewicz, N. Paar, et al., Phys. Rev. **C85**, 041302(R) (2012).
[55] A. Tamii et al., Phys. Rev. Lett. **107**, 062502 (2011).
[56] I. Poltoratska, P. von Neumann-Cosel, A. Tamii, T. Adachi, C. Bertulani, et al. (2012), `arXiv:1203.2155`.
[57] P.-G. Reinhard and W. Nazarewicz, Phys. Rev. **C81**, 051303 (2010).
[58] F. Fattoyev and J. Piekarewicz, Phys. Rev. **C84**, 064302 (2011).
[59] F. Fattoyev and J. Piekarewicz, Phys. Rev. **C88**, 015802 (2012).
[60] Y. Suzuki, K. Ikeda, and H. Sato, Prog. Theor. Phys. **83**, 180 (1990).
[61] P. Van Isacker and D. D. Nagarajan, M. A.and Warner, Phys. Rev. **C45**, R13 (1992).
[62] I. Hamamoto, H. Sagawa, and X. Z. Zhang, Phys. Rev. **C53**, 765 (1996).
[63] I. Hamamoto, H. Sagawa, and X. Z. Zhang, Phys. Rev. **C57**, R1064 (1998).
[64] D. Vretenar, N. Paar, P. Ring, and G. A. Lalazissis, Phys. Rev. **C63**, 047301 (2001a).
[65] D. Vretenar, N. Paar, P. Ring, and G. A. Lalazissis, Nucl. Phys. **A692**, 496 (2001b).
[66] N. Paar, T. Niksic, D. Vretenar, and P. Ring, Phys. Lett. **B606**, 288 (2005).
[67] J. Piekarewicz, Phys. Rev. **C73**, 044325 (2006).

Journal of Physics: Conference Series **492** (2014) 012009 doi:10.1088/1742-6596/492/1/012009

Algebraic treatment of alpha-cluster nuclei

Roelof Bijker

Instituto de Ciencias Nucleares, Universidad Nacional Autónoma de México, A.P. 70-543, 04510 México, D.F., México

E-mail: bijker@nucleares.unam.mx

Abstract. I discuss an algebraic treatment of alpha-cluster nuclei in the framework of the Algebraic Cluster Model. It is suggested to treat ^{12}C as an oblate top with $\mathcal{D}_{3h}$ symmetry (equilateral triangle) and ^{16}O as a spherical top with tetrahedral $\mathcal{T}_d$ symmetry (regular tetrahedron). The structure of the rotational bands is a consequence of the underlying symmetry, and can be used to help determine the geometrical configuration of the α-particles.

1. Introduction

The structure of ^{12}C has been the subject of many studies, both experimental and theoretical, since the early days of nuclear physics [1]. Especially the (collective) nature of the 0^+ Hoyle state at 7.65 MeV which is of crucial importance in stellar nucleosynthesis to explain the observed abundance of ^{12}C, has presented a formidable challenge to nuclear structure calculations, such as cluster model [2], (no-core) shell model [3, 4] or *ab initio* lattice calculations [5].

Recently, three different experimental collaborations reported the measurement for a lowlying 2^+ state around 10 MeV with a width of approximately 0.8 MeV which was interpreted as a rotational excitation of the Hoyle state, thus providing the first evidence of a collective band structure built on top of the Hoyle state. This 2^+ state was observed in the $^{12}\mathrm{C}(\alpha,\alpha')$ reaction at an energy 9.75 ± 0.15 MeV and width 0.75 ± 0.15 MeV [6]. A combination of the (α,α') and (p,p') reactions showed evidence for a 2^+ state at $E = 9.75 \pm 0.15$ MeV and $\Gamma = 0.75 \pm 0.15$ MeV [7]. Finally, experiments using the $^{12}\mathrm{C}(\gamma,\alpha)^8\mathrm{Be}$ reaction showed a state at 10.03 ± 0.11 MeV and $\Gamma = 0.80 \pm 0.13$ MeV [8].

In view of the new experimental information, in this contribution I revisit an algebraic treatment of α-cluster nuclei, called the Algebraic Cluster Model, and discuss two special solutions which are relevant to the α-cluster nuclei ^{12}C [9] and ^{16}O [10].

2. Algebraic Cluster Model

Algebraic models and spectrum generating algebras have played an important role in the study of both many-body and few-body systems. Generally speaking, in algebraic models energy eigenvalues and eigenvectors are obtained by diagonalizing a finite-dimensional matrix, rather than by solving a set of coupled differential equations in coordinate space. As examples, I mention the interacting boson model which has been very successful in the description of collective states in nuclei [11], and the vibron model [12] which was introduced to describe vibrational and rotational excitations in diatomic molecules. In spectroscopic studies these algebraic methods provide a powerful tool to study symmetries and selection rules, to classify the basis states, and to calculate matrix elements of physical observables.

Journal of Physics: Conference Series **492** (2014) 012009 doi:10.1088/1742-6596/492/1/012009

The Algebraic Cluster Model (ACM) is an interacting boson model to describe the relative motion of the n clusters based on the spectrum generating algebra of $U(\nu+1)$ where $\nu = 3(n-1)$ represents the number of relative spatial degrees of freedom (after removal of the center of mass). For the two-body problem the ACM reduces to the $U(4)$ vibron model [12] and for three-body clusters to the $U(7)$ model [9, 13, 14, 15].

The spatial degrees of freedom for n-body systems are taken as the relative Jacobi coordinates $\vec{\rho}_k$ and the center of mass $\vec{R}$

$$\begin{aligned} \vec{\rho}_k &= \frac{1}{\sqrt{k(k+1)}}\left(\sum_{i=1}^{k} \vec{r}_i - k\vec{r}_{k+1}\right) , \qquad k = 1,\ldots,n-1 , \\ \vec{R} &= \frac{1}{\sqrt{n}}\sum_{i=1}^{n} \vec{r}_i , \end{aligned} \qquad (1)$$

and their conjugate momenta. Here $\vec{r}_i$ are the coordinates of the i-th cluster $(i = 1,\ldots,n)$. Instead of a formulation in terms of coordinates and momenta the method of bosonic quantization is used which consists of introducing a dipole boson with $L^P = 1^-$ $(b_k^\dagger)$ for each independent relative Jacobi vector $(k = 1,\ldots,n-1)$, and an auxiliary scalar boson with $L^P = 0^+$ $(s^\dagger)$. The scalar boson does not represent an independent degree of freedom, but is added under the restriction that the total number of bosons $N = n_s + \sum_k n_k$ is conserved. This procedure leads to a compact spectrum generating algebra of $U(3n-2)$ whose model space is spanned by the symmetric irreducible representation $[N]$ which contains the oscillator shells with $n_b = \sum_k n_k = 0,1,2,\ldots,N$. The introduction of the scalar boson makes it possible to investigate the dynamics of $n-1$ vector degrees of freedom including situations in which there is a mixing of oscillator shells.

In the present application to α-cluster nuclei the Hamiltonian has to be invariant under the permuation group S_n for n identical clusters. As a result, the most general one- and two-body Hamiltonian that describes the relative motion of a system of n identical clusters, is a scalar under the permutation group S_n, is rotationally invariant, conserves parity as well as the total number of bosons N, is given by

$$\begin{aligned} H &= \epsilon_0\, s^\dagger\tilde{s} - \epsilon_1 \sum_k b_k^\dagger\cdot\tilde{b}_k + u_0\, s^\dagger s^\dagger\,\tilde{s}\tilde{s} - u_1 \sum_k s^\dagger b_k^\dagger\cdot\tilde{b}_k\tilde{s} \\ &\quad + v_0\left[\sum_k b_k^\dagger\cdot b_k^\dagger\,\tilde{s}\tilde{s} + \text{h.c.}\right] + \sum_L \sum_{ijkl} v_{ijkl}^{(L)}\, [b_i^\dagger \times b_j^\dagger]^{(L)}\cdot[\tilde{b}_k \times \tilde{b}_l]^{(L)} , \end{aligned} \qquad (2)$$

with $\tilde{b}_{k,m} = (-1)^{1-m} b_{k,-m}$ and $\tilde{s} = s$. The invariance under the permutation symmetry imposes further restrictions on the coefficients $v_{ijkl}^{(L)}$. By construction, the first five terms in Eq. (2) are invariant under the permutation group S_n.

In general, the eigenvalues and corresponding eigenvectors are obtained numerically by diagonalizing the Hamiltonian in an appropriate basis. By construction, the wave functions are characterized by the total number of bosons N, angular momentum and parity L^P, and their transformation property under t the permutation group S_n. Since in the application to α-cluster nuclei the excitation of the α particles themselves is not considered, the wave functions describing the relative motion have to be symmetric under S_n.

The ACM has a rich algebraic structure, which includes both continuous and discrete symmetries. It is of general interest to study limiting cases of the Hamiltonian in which the energy spectra can be obtained in closed form. The S_n invariant ACM Hamiltonian of Eq. (2)

has two dynamic symmetries corresponding to the group chains

$$U(3n-2) \;\supset\; \left\{ \begin{array}{l} U(3n-3) \\ \\ SO(3n-2) \end{array} \right. \tag{3}$$

These dynamic symmetries were studied in general for the n-body problem and were shown to correspond to the $(3n-3)$-dimensional (an)harmonic oscillator and a deformed oscillator, respectively [10].

In this contribution, I review two other limiting cases of S_n invariant Hamiltonians for which approximate solutions can be obtained in a semiclassical mean-field analysis and which could be of interest for the application to α-cluster nuclei. For the case of three-body clusters I discuss the oblate symmetric top that corresponds to three particles located at the vertices of an equilateral triangle, and for four-body clusters the spherical top with four particles at the vertices of a regular tetrahedron.

3. Three-body clusters: oblate top

For the case of three identical clusters the spectrum generating algebra of the ACM is given by $U(7)$. Here I consider a Hamiltonian which is special case of Eq. (2)

$$\begin{aligned} H &= \xi_1 \, (R^2 \, s^\dagger s^\dagger - b_1^\dagger \cdot b_1^\dagger - b_2^\dagger \cdot b_2^\dagger) \, (R^2 \, \tilde{s}\tilde{s} - \tilde{b}_1 \cdot \tilde{b}_1 - \tilde{b}_2 \cdot \tilde{b}_2) \\ & + \xi_2 \left[(b_1^\dagger \cdot b_1^\dagger - b_2^\dagger \cdot b_2^\dagger) \, (\tilde{b}_1 \cdot \tilde{b}_1 - \tilde{b}_2 \cdot \tilde{b}_2) + 4 \, (b_1^\dagger \cdot b_2^\dagger) \, (\tilde{b}_2 \cdot \tilde{b}_1) \right] \\ & + 2\kappa_1 \, (b_1^\dagger \times \tilde{b}_1 + b_2^\dagger \times \tilde{b}_2)^{(1)} \cdot (b_1^\dagger \times \tilde{b}_1 + b_2^\dagger \times \tilde{b}_2)^{(1)} \\ & + 3\kappa_2 \, (b_1^\dagger \times \tilde{b}_2 - b_2^\dagger \times \tilde{b}_1)^{(0)} \cdot (b_2^\dagger \times \tilde{b}_1 - b_1^\dagger \times \tilde{b}_2)^{(0)} \; . \end{aligned} \tag{4}$$

For $R^2 = 0$, this Hamiltonian has $U(7) \supset U(6)$ symmetry (anharmonic oscillator), whereas for $R^2 = 1$ and $\xi_2 = 0$ it has $U(7) \supset SO(7)$ symmetry (deformed oscillator). The general case with $R^2 \neq 0$ and ξ_1, $\xi_2 > 0$ corresponds to a potential energy surface with the equilibrium shape of an equilateral triangle, in which the two Jacobi vectors $\vec{\rho}_1$ and $\vec{\rho}_2$ have equal length and are perpendicular [9]. The corresponding point group symmetry is $\mathcal{D}_{3h}$ whose subgroup D_3 is isomorphic to the permutation group S_3. Even though in this case the energy eigenvalues cannot be derived in closed form, an approximate expression for the energy levels can be obtained by making use of the method of intrinsic or coherent states (valid in the limit of large N) [13, 15]

$$E(v_1, v_2^l, L, K) = \omega_1 \, (v_1 + \frac{1}{2}) + \omega_2 \, (v_2 + 1) + \kappa_1 \, L(L+1) + \kappa_2 \, (K \pm 2l)^2 \; , \tag{5}$$

with frequencies $\omega_1 = 4NR^2\xi_1$ and $\omega_2 = 4NR^2\xi_2/(1+R^2)$. The quantum numbers have the following meaning: v_1, v_2 are vibrational quantum numbers: v_1 corresponds to a symmetric one-dimensional stretching vibration, and v_2 represents a two-dimensional bending vibration; $l = v_2, v_2-2, \ldots, 1$ or 0 is the vibrational angular momentum of the doubly degenerate vibration; L is the angular momentum, and K its projection on a body-fixed axis [9].

Fig. 1 shows the rotation-vibration spectrum of a triangular configuration with $\mathcal{D}_{3h}$ symmetry which is characterized by a series of rotational bands labeled by (v_1, v_2^l). For this case, the bands with $(v_1, 0^0)$ can have angular momenta and parity $L^P = 0^+, 2^+, 3^-, 4^\pm, \ldots$, whereas the angular momentum content of the doubly degenerate vibrations $(v_1, 1^1)$ is given by $L^P = 1^-, 2^\mp, 3^\mp, \ldots$, in agreement with Ref. [16]. The structure of the rotational bands can be considered as the fingerprint of the underlying $\mathcal{D}_{3h}$ symmetry of three identical structureless particles at the vertices of an equilateral triangle.

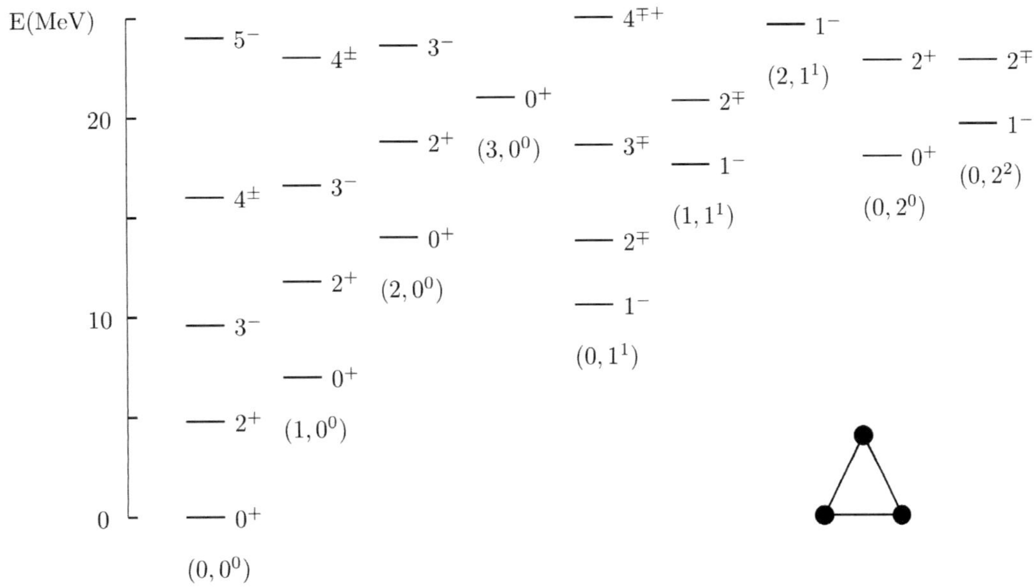

Figure 1. Rotation-vibration spectrum of an oblate top with $\mathcal{D}_{3h}$ symmetry calculated using Eq. (5) with $\omega_1 = 7.0$, $\omega_2 = 9.0$, $\kappa_1 = 0.8$ and $\kappa_2 = 0.0$ MeV (only the levels with $E \leq 25$ MeV are shown). The levels are characterized by angular momentum and parity L^P, and the vibrational labels (v_1, v_2^l). All states are symmetric under $S_3 \sim D_3$.

The nucleus ^{12}C

The structure of ^{12}C has been extensively investigated in α-cluster models [1, 2, 17], the shell model from early calculations in the p-shell [18] to more recent studies with large shell mixing in the *ab initio* [3] or symmetry-adapted no-core shell model [4, 19], as well as *ab initio* calculations based on lattice effective field theory [5, 20].

In this contribution, the spectrum of ^{12}C is analyzed in the oblate top limit of the ACM which corresponds to a triangular configuration in which the three α particles are located at the vertices of an equilateral triangle (point group $\mathcal{D}_{3h}$). The experimental spectrum of ^{12}C is shown in Fig. 2, where it is compared with the exact energy spectrum of the oblate top Hamiltonian of Eq. (4) with $N = 10$ bosons [9]. The coefficients ξ_1, ξ_2, κ_1 and κ_2 were determined in a fit to the excitation energies of ^{12}C (see Table 1). One can clearly identify in the experimental spectrum the states 0^+, 2^+, 3^-, 4^+ of the ground-state rotational band, and possibly the 4^- state at 13.35 MeV [21]. The 3^- state does not fall at the location expected for an oblate top, but at a somewhat higher excitation energy. Since the 3^- state has $K = 3$ the deviation from the simple rotational formula indicates large rotation-vibration interactions. The spectrum also shows an excited 0^+ state at 7.65 MeV and a 1^- state at 10.84 MeV which could be interpreted as bandheads of the vibrational (stretching and doubly degenerate bending) excitations. In order to determine whether or not this is the case or rather those states represent other types of configurations, such as three α particles on a line [22] or a bent-arm configuration as suggested

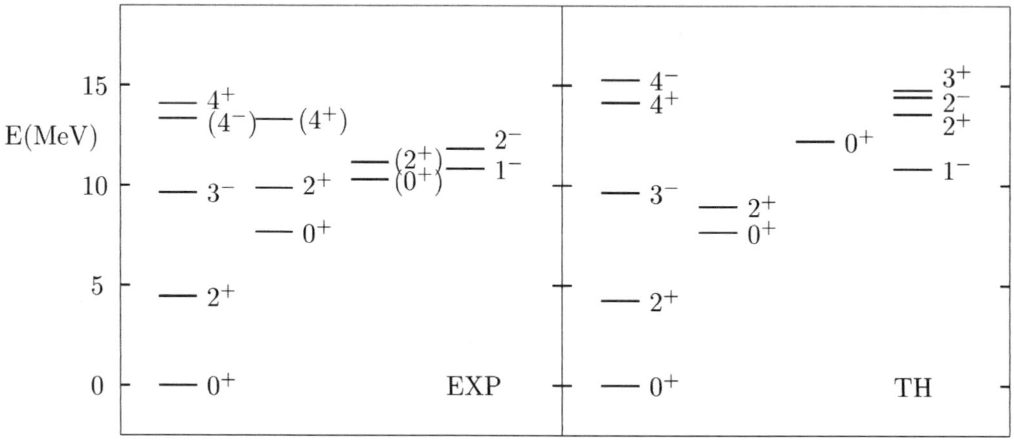

Figure 2. Comparison between the low-lying experimental spectrum of ^{12}C and that calculated with the oblate top Hamiltonian of Eq. (4) with $N = 10$. The parameter values are given in Table 1. States with uncertain spin-parity assignment are in parentheses.

recently in *ab initio* lattice calcualtions [5], one would have to identify the rotational sequences built on top of them which have a characteristic pattern for triangular configurations and another pattern for linear or bent configurations. In this respect, it is of special interest to investigate in more detail the nature of the 2^+ state at 11.16 MeV and 2^- state at 11.83 MeV, which in the oblate top model could be interpreted as rotational excitations of the doubly degenerate vibration.

Recently, the structure of the Hoyle band was studied experimentally with the measurement of a new 2^+ state with an energy in the range 9.75-10.03 MeV and a width of the order of 0.75-1.01 MeV [6, 7, 8], which was interpreted as a possible candidate for a rotational excitation of the 0^+ Hoyle state. In Fig. 2, this new state is included by taking the average of the three measured energies. This assignment is supported by a recent analysis of the rms radii of the 0_2^+ and 2_2^+ states [23]. The radii were found to be quite similar thus providing evidence that these states belong to the same rotational band. In addition, there is experimental evidence for a state at 13.3 MeV which was assigned tentatively as a 4^+ state belonging to the Hoyle band [24]. In order to distinguish between different geometrical configurations for the Hoyle band, such as linear, bent or triangular (see Fig. 3), the measurement of a possible 3^- Hoyle state is

Table 1. Coefficients used in the calculation of the energy spectrum and the transition form factors of ^{12}C. The number of bosons is $N = 10$.

ξ_1	0.1721	MeV	κ_1	0.7068	MeV
ξ_2	0.2745	MeV	κ_2	0.1276	MeV
R^2	1.40				
β	1.74	fm	α	0.52	fm^{-2}

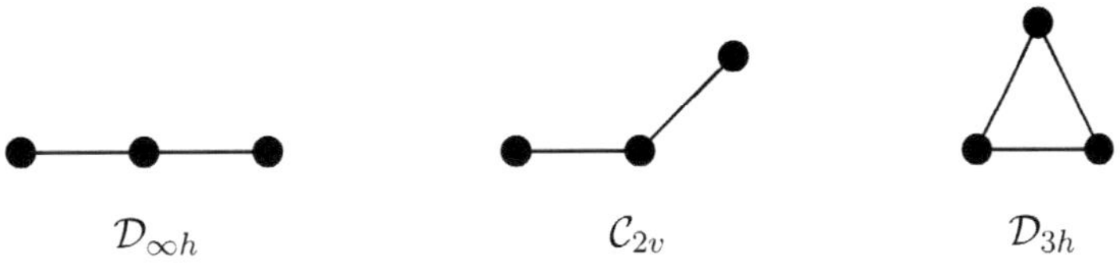

Figure 3. Different α-cluster configurations for ^{12}C: linear, bent and triangular with their respective point-group symmetries.

crucial, since its presence would indicate a triangular configuration, just as for the ground state band.

The moments of inertia of the ground band and the Hoyle band are very different. This leads to a somewhat larger radius for the Hoyle band than for the ground state band. Since for the oblate top, both the ground-state band and the Hoyle band arise from the same geometrical shape of an equilateral triangle, the radius of the Hoyle state can be estimated from the moments of inertia to be a factor $1.4 - 1.5$ (depending on the energy of the 2^+ Hoyle state [6, 7, 8]) larger than the rms radius of the ground state [9]. This value is a bit higher than, but essentially consistent with the one determined in the modified diffraction model [23].

For the oblate top, electromagnetic transition rates and form factors can be obtained in explicit form in the limit of large N. For example, the elastic form factor is given by .

$$F(0_1^+ \rightarrow 0_1^+; q) \quad = \quad j_0(q\beta)\, \mathrm{e}^{-q^2/4\alpha} \; . \tag{6}$$

The coefficient β is determined from the first minimum in the elastic form factor, and subsequently the coefficient α is determined from the charge radius of ^{12}C (see Table 1). Finally, since the energy spectrum is not very sensitive on the coefficient R^2 in the first term of the Hamiltonian of Eq. (4), its value is determined from the first minimum in the transition form factor to the Hoyle state $F(0_1^+ \rightarrow 0_2^+; q)$ [9].

Fig. 4 shows a comparison between experimental and theoretical form factors. The calculations give a reasonable description of the transition form factors to the rotational excitations of the ground state band, as well as the shape of the form factors leading to the 0^+ Hoyle state, indicating that the 0_2^+ state could indeed be the vibrational excitation of a three-alpha configuration with $\mathcal{D}_{3h}$ symmetry albeit with a large mixing with other configurations. Finally, the transition form factor for the 2_2^+ state shows a similar structure as the 0_2^+ state.

4. Four-body clusters: spherical top

As a second example, I discuss the spherical top with tetrahedral symmetry $\mathcal{T}_d$ as a special case of the ACM for four identical clusters. The spectrum generating algebra is given by $U(10)$. Let us consider a S_4 invariant Hamiltonian of the form

$$\begin{aligned}
H_4 \;=\; & \xi_1\,(R^2\, s^\dagger s^\dagger - b_1^\dagger \cdot b_1^\dagger - b_2^\dagger \cdot b_2^\dagger - b_3^\dagger \cdot b_3^\dagger)\,(\mathrm{h.c.}) \\
& + \xi_2 \Big[(-2\sqrt{2}\, b_1^\dagger \cdot b_3^\dagger + 2 b_1^\dagger \cdot b_2^\dagger)\,(\mathrm{h.c.}) \\
& \qquad + (-2\sqrt{2}\, b_2^\dagger \cdot b_3^\dagger + (b_1^\dagger \cdot b_1^\dagger - b_2^\dagger \cdot b_2^\dagger))\,(\mathrm{h.c.}) \Big] \\
& + \xi_3 \Big[(2 b_1^\dagger \cdot b_3^\dagger + 2\sqrt{2}\, b_1^\dagger \cdot b_2^\dagger)\,(\mathrm{h.c.})
\end{aligned}$$

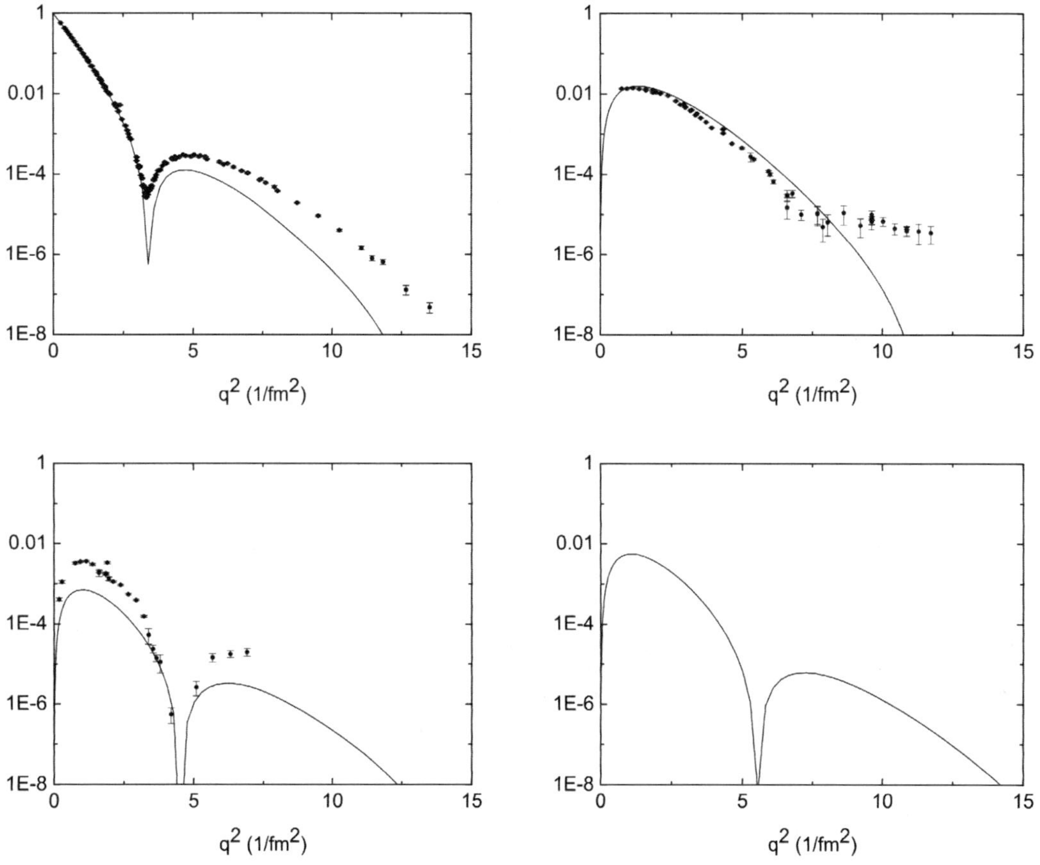

Figure 4. Comparison between the experimental form factors $|\mathcal{F}(0_1^+ \to L_i^P; q)|^2$ of ^{12}C for the final states $L_i^P = 0_1^+$ (top left), 2_1^+ (top right), 0_2^+ (bottom left) and 2_2^+ (bottom right) and those obtained for the oblate top with $N = 10$. The parameter values are given in Table 1. The experimental data are taken from [25]-[30].

$$
\begin{aligned}
&+(2b_2^\dagger \cdot b_3^\dagger + \sqrt{2}\,(b_1^\dagger \cdot b_1^\dagger - b_2^\dagger \cdot b_2^\dagger))\,(\text{h.c.}) \\
&+(b_1^\dagger \cdot b_1^\dagger + b_2^\dagger \cdot b_2^\dagger - 2b_3^\dagger \cdot b_3^\dagger)\,(\text{h.c.})\Big] \\
+\kappa_1\,\vec{L}\cdot\vec{L} + \kappa_2\,(\vec{L}\cdot\vec{L} - \vec{I}\cdot\vec{I})^2 \;, \qquad (7)
\end{aligned}
$$

where $\vec{L}$ denotes the angular momentum in coordinate space and $\vec{I}$ the angular momentum in index space. For $R^2 = 0$, this Hamiltonian has $U(10) \supset U(9)$ symmetry (anharmonic oscillator), whereas for $R^2 = 1$ and $\xi_2 = \xi_3 = 0$ it has $U(10) \supset SO(10)$ symmetry (deformed oscillator). The general case with $R^2 \neq 0$ and ξ_1, ξ_2, $\xi_3 > 0$ corresponds to a geometric configuration in which the four clusters are located at the vertices of a regular tetrahedron: *i.e.* the three Jacobi vectors $\vec{\rho}_1$, $\vec{\rho}_2$ and $\vec{\rho}_3$ have equal length and are mutually perpendicular [10]. The corresponding point group symmetry is $\mathcal{T}_d$ which is isomorphic to the permutation group S_4. Just as in the previous case of the oblate top, also for the Hamiltonian of Eq. (7) an approximate energy formula can be derived in a semiclassical mean-field analysis by making use of the methods of intrinsic or coherent states which is valid in the limit of large N. The resulting energy spectrum is that of the vibrational and rotational excitations of a spherical top with tetrahedral symmetry

$$E = \omega_1(v_1 + \frac{1}{2}) + \omega_2(v_2 + 1) + \omega_3(v_3 + \frac{3}{2}) + \kappa_1 L(L+1) , \tag{8}$$

with frequencies $\omega_1 = 4NR^2\xi_1$, $\omega_2 = 8NR^2\xi_2/(1+R^2)$ and $\omega_3 = 8NR^2\xi_3/(1+R^2)$. Here v_1 represents the vibrational quantum number for a symmetric stretching A vibration, v_2 denotes a doubly degenerate E vibration, and v_3 a three-fold degenerate F vibration.

Whereas the angular momentum L is an exact symmetry of H_4 of Eq. (7), the angular momentum in index space I in general does not commute with the Hamiltonian. Only if $\xi_2 = \xi_3$ in Eq. (7), does I become a good quantum number. The rotational excitations of the ground state vibrational band of H_4 with $(v_1, v_2, v_3) = (0, 0, 0)$ are characterized by $L = I$. This property is a consequence from the fact that the operator $\vec{L} \cdot \vec{L} - \vec{I} \cdot \vec{I}$ annihilates the coherent (or intrinisic) state corresponding to the rigid equilibrium shape of a regular tetrahedron.

Fig. 5 shows the structure of the spectrum of a spherical top with tetrahedral symmetry according to the approximate energy formula of Eq. (8). The energy spectrum consists of a series of rotational bands labeled by (v_1, v_2, v_3). It is assumed that the spin of the identical clusters is zero, as is relevant for the description of the ^{16}O nucleus as a cluster of four α particles. As a consequence, all states in Fig. 5 are symmetric under S_4. For this case, the bands with $(v_1, 0, 0)$ can have angular momenta and parity $L^P = 0^+, 3^-, 4^+, 6^\pm, \ldots$, whereas the angular momentum content of the doubly degenerate E vibration $(v_1, 1, 0)$ is given by $L^P = 2^\pm, 4^\pm, 5^\pm, \ldots$, and for the triply degenerate F vibration $(v_1, 0, 1)$ by $L^P = 1^-, 2^+, 3^\pm, 4^\pm, \ldots$, in agreement with Ref. [16]. A more detailed analysis of ^{16}O in the ACM is forthcoming.

5. Summary and conclusions

In this contribution, I presented an algebraic treatment of α-cluster nuclei in terms of the Algebraic Cluster Model which is based on the algebraic quantization of the relative Jacobi variables. The ensuing $U(3n-2)$ spectrum generating algebra for a system on n clusters incorporates all vibrational and rotational degrees of freedom from the outset. The permutation symmetry of the α-clusters is taken into account in an exact manner. Two cases were discussed in more detail.

First it was suggested to describe the nucleus ^{12}C as a system of three α-particles located at the vertices of an equilateral triangle (oblate top). The spectrum of the ground state band of a triangular configuration is characterized by the rotational sequence: 0^+, 2^+, 3^-, $4^\pm$, ..., as is indeed observed in ^{12}C. A linear configuration would not have negative parity states, while a shell-model configuration would not have the 3^- state as a member of the rotational band but rather as an octupole vibration, *i.e.* it would not form a rotational sequence with the 0^+, 2^+, 4^+ states. For this reason it is very important to look for the 3^- state of the Hoyle band in order to determine whether the geometrical configuration of the α-particles for the Hoyle band is linear, bent or triangular.

As a second example, I showed some results for four identical clusters located at the vertices of a regular tetrahedron (spherical top) with possible applications for the nucleus ^{16}O as a system of four α-particles. As a consequence of the tetrahedral symmetry of this configuration, the ground state rotational band is given by the sequence 0^+, 3^-, 4^+, $6^\pm$, ..., just as is observed in the nucleus ^{16}O. The absence of a 2^+ state is due to the tetrahedral symmetry.

It is important to stress that these rotational sequences can be considered as the fingerprints of the underlying point-group symmetry of the system. If a physical system consists of three identical particles at the vertices of an equilateral triangle, then its spectrum *must* be as in Fig. 1, or for a system composed of four identical particles at the vertices of a regular tetrahedron, the spectrum *must* be as in Fig. 5.

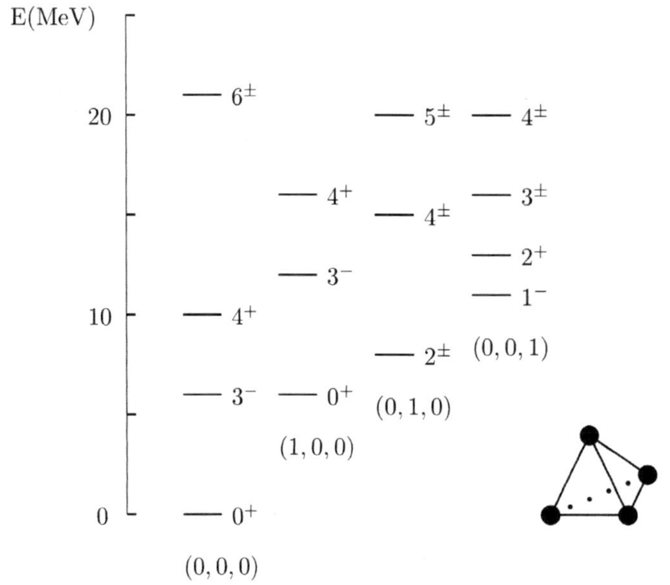

Figure 5. Rotation-vibration spectrum of a spherical top with tetrahedral $\mathcal{T}_d$ symmetry calculated using Eq. (8) with $\omega_1 = 6.0$, $\omega_2 = 5.0$, $\omega_3 = 10.0$ and $\kappa = 0.5$ MeV (only the levels belonging to the bands with $v_1 + v_2 + v_3 \leq 1$ are shown). The levels are characterized by angular momentum and parity L^P, and the vibrational labels (v_1, v_2, v_3). All states are symmetric under $S_4 \sim \mathcal{T}_d$.

In contrast with molecules, the vibrational and rotational frequencies in nuclei are comparable which implies large rotation-vibration interactions. The finite value of the number of bosons N implies that the vibrational bands are no longer decoupled, but instead show an appreciable mixing between them.

Finally, it is important to note that the ACM provides a general framework to study the full rotational and vibrational structure of cluster systems which is not restricted to the case of identical particles discussed in this contribution. It can be applied to other situations as well, such as nonidentical particles [15] or other geometric configurations, rigid as well as non-rigid.

Acknowledgments

This work was supported in part by CONACyT (grant 78833) and and PAPIIT-UNAM (grant IN113711).

References

[1] Wheeler J A 1937 *Phys. Rev.* **52** 1083
[2] Von Oertzen W, Freer M and Kanada-En'yo Y 2006 *Phys. Rep.* **432** 43
[3] Roth R, Langhammer J, Calci A, Binder S and Navrtil P, *Phys. Rev. Lett.* **107** 072501
[4] Dreyfuss A C, Launey K D, Draayer J P, Dytrych T, and Bahri C 2012 *AIP Conf Proc* **1488** 204
[5] Epelbaum E, Krebs H, Lähde T, Lee D and Meissner U G 2012 *Phys. Rev. Lett.* **109** 252501

[6] Itoh M, *et al.* 2011 *Phys. Rev. C* **84** 054308
[7] Freer M, *et al.* 2012 *Phys. Rev. C* **86** 034320
[8] Zimmerman W R, *et al.* 2013 *Phys. Rev. Lett.* **110** 152502
[9] Bijker R and Iachello F 2000 *Phys. Rev. C* **61** 067305
Bijker R and Iachello F 2002 *Ann. Phys. (N.Y.)* **298** 334
[10] Bijker R 2010 *AIP Conference Proceedings* **1323** 28
Bijker R 2012 *J. Phys.: Conf. Ser.* **380** 012003
[11] Iachello F and Arima A 1987 *The Interacting Boson Model* (Cambridge: Cambridge U. Press)
[12] Iachello F and Levine R D 1995 *Algebraic Theory of Molecules* (Oxford: Oxford U. Press)
[13] Bijker R, Iachello F and Leviatan A 1994 *Ann. Phys. (N.Y.)* **236** 69
Bijker R, Iachello F and Leviatan A 2000 *Ann. Phys. (N.Y.)* **284** 89
[14] Bijker R, Dieperink A E L and Leviatan A 1995 *Phys. Rev. A* **52** 2786
[15] Bijker R and Leviatan A 1998 *Few-Body Systems* **25** 89
[16] Herzberg G 1991 *Molecular Spectra and Molecular Structure. II. Infrared and Raman Spectra of Polyatomic Molecules* (Malabar Florida: Krieger)
[17] Robson D 1978 *Nucl. Phys. A* **308** 381
[18] Cohen S and Kurath D 1965 *Nucl. Phys.* **73** 1
[19] Dreyfuss A C, Launey K D, Dytrych T, Draayer J P and Bahri C 2012 arXiv:1212.2255 [nucl-th]
[20] Epelbaum E, Krebs H, Lee D and Meissner U G 2011 *Phys. Rev. Lett.* **106** 192501
[21] Freer M *et al.* 2007 *Phys. Rev. C* **76** 034320
[22] Morinaga H 1956 *Phys. Rev.* **101** 254
[23] Danilov A N, Belyaeva T L, Demyanova A S, Goncharov S A and Ogloblin A A 2009 *Phys. Rev. C* **80** 054603
Ogloblin A A, Belyaeva T L, Danilov A N, Demyanova A S and Goncharov S A 2013 *Eur. Phys. J. A* **49** 46
[24] Freer M, *et al.* 2012 *Phys. Rev. C* **83** 034314
[25] Reuter W, Fricke G, Merle K and Miska H 1982 *Phys. Rev. C* **26** 806
[26] Sick I and McCarthy J S 1970 *Nucl. Phys. A* **150** 631
[27] Crannell H L and Griffy T A 1964 *Phys. Rev.* **136** B1580
[28] Crannell H 1966 *Phys. Rev.* **148** 1107
[29] Crannell H, O'Brien J T and Stober D I 1979 *Int. Conf. on Nuclear Physics with Electromagnetic Interactions*
[30] Strehl P and Schucan Th H 1968 *Phys. Lett. B* **27** 641

Journal of Physics: Conference Series **492** (2014) 012010 doi:10.1088/1742-6596/492/1/012010

Nuclear physics experiments with a windowless supersonic gas jet target

J F Favela[1], D Shapira[2], E Chávez[1], M E Ortíz[1], E Andrade[1], O G de Lucio[1] and A Huerta[1]

[1] Instituto de Física, Universidad Nacional Autónoma de México, México, D.F. 04510, Mexico
[2] Oak Ridge National Laboratory, Oak Ridge, TN 37831, USA

E-mail: ffavela@gmail.com

Abstract.
A new windowless gas target has been developed in Mexico. It is a supersonic gas jet flow produced inside a vacuum chamber which can be coupled to a regular beam line in an accelerator laboratory as a differential pumping system brings the pressure of the gas target system down to a microTorr, or better, at the connecting stage.

In this work, we present the system as it was designed and constructed as well as the first results using air, Nitrogen and Argon.

1. Introduction

Nuclear data (reaction dynamics, structure) becomes harder to obtain when experiments have to be performed with accelerated particles impinging on nuclei that are found in nature as gases. Sometimes it is possible to produce a solid compound that would contain the nucleus of interest (Hydrides for instances to study reactions on H), or build gas cells with thin windows for noble gases. These and other strategies may not always be appropriate because of the interference of the other materials present in the compound or the windows. It can also be unappropiate because the target may have low density as in the case of windowless gas targets. Gas jets are an optimal solution. In the present work a new facility developed for the Carlos Graef Fernández Laboratory of the Physics Institute at UNAM (CGFL) is described in detail. The CGFL is a facility powered by a CN-Van de Graaff accelerator. This machine provides light ion beams with energies up to $5.5MeV$, ideal for low energy nuclear physics studies. First data on different jets (Air, N, Ar) is presented.

A high density and spatially well defined gas film constitutes an ideal target for low count rate nuclear physics experiments, since they can sustain several orders of magnitude more beam intensity than a solid target without being altered. Such a gas target can be produced by a supersonic jet injected in the reaction chamber in vacuum.

Nuclear studies can be carried out on some gas substances by producing a chemical compound that is solid. Such is the case, for instances, of studies on Hydrogen, where it appears in the target as part of a molecule. In those cases, high intensity beams can change the Hydrogen content in the target as a function of time, and eventually destroy it. Also beam interactions with the other components of the molecules have to be taken into account.

Some gases cannot form molecules (noble), gas cells have been used to contain a certain amount of gas in the path of the beam. There the solid windows used to isolate the gas from the vacuum of the accelerator beam line limits the beam intensity as well.

The supersonic gas jet strategy to produce a target of a material that is found as gas in nature to be bombarded with accelerated particles is not new. An example of such devices made in the past can be found here [1]. An example of modern supersonic gas jet systems has just been finished at the University of Notre Dame in the USA [2]. The needed general theory can be found in [3]and [4].

In section 2, our device is presented in detail. Section 3 shows the results of the characterization of the windowless supersonic gas jet target. Some conclusions can be found in section 4.

2. Experimental Methods

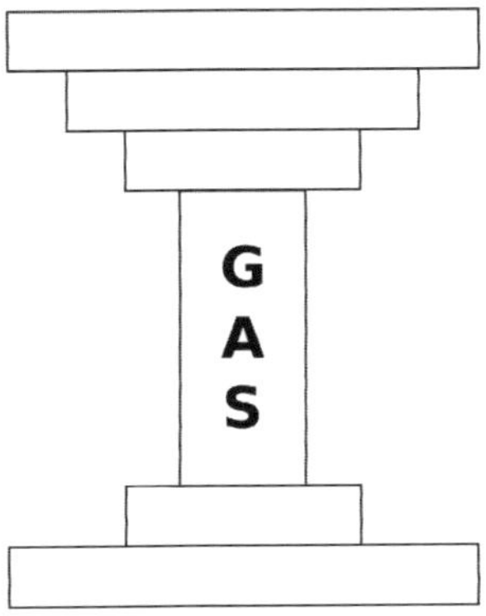

(a)

(b)

Figure 1. (a) Shows an artistic representation of the gas jet: a high density gas column with a well defined frontier surrounded by high vacuum, (b) shows the jet target getting hit by the accelerator's beam, notice the faint purple glow. The image is superimposed with part of the experimental array for better reference.

In figure 1 (a), a representation of the jet target is shown as a column of gas without walls. Also shown in the figure (b) a photograph of the accelerator's beam hitting the jet target, which causes the jet area to glow. In this experiment $450keV$ protons where used with a current of 1.5 μA. More details are also shown, the gas jet comes out of a rectangular nozzle (top) and is immediately caught by a pipe (bottom) connected to a powerful vacuum pump. Both the nozzle and the pipe are mounted in a way where they can rotate and their relative height can be adjusted.

For the experiments reported here a rectangular shape was chosen for the nozzle. The gas jet produced resembles a planar target and we can be adjust its density by simply rotating it. Nozzles of different shape are available and are easy to change.

In order to characterize the properties of our gas jet, experiments were made with air, Nitrogen and Argon. Gases that are non corrosive, toxic or explosive. Argon and Nitrogen are also nearly mono-isotopic elements in nature.

In figure 2, the experimental setup inside the scattering chamber used for the characterization of the target using Rutherford Back Scattering is shown schematically. The rectangular boxes represent charged particle detectors, one of which is shown in the superimposed image, figure 1 (b).

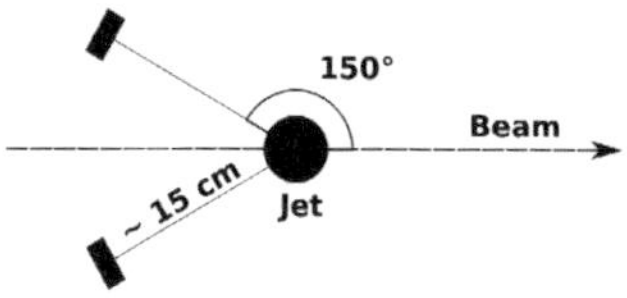

Figure 2. RBS setup.

Gas is injected into the system at high pressures: 1-5 atmospheres. The pressure in the region surrounding the jet is of the order of 0.1 Torr meaning that it is too high for connecting to the accelerator's beam line that is kept at high vacuum (10^{-6} Torr or better). In order to connect the system a differential pumping system is needed, lowering gradually the pressure from the scattering chamber to the accelerator. In figure 3 an outline of the system is shown.

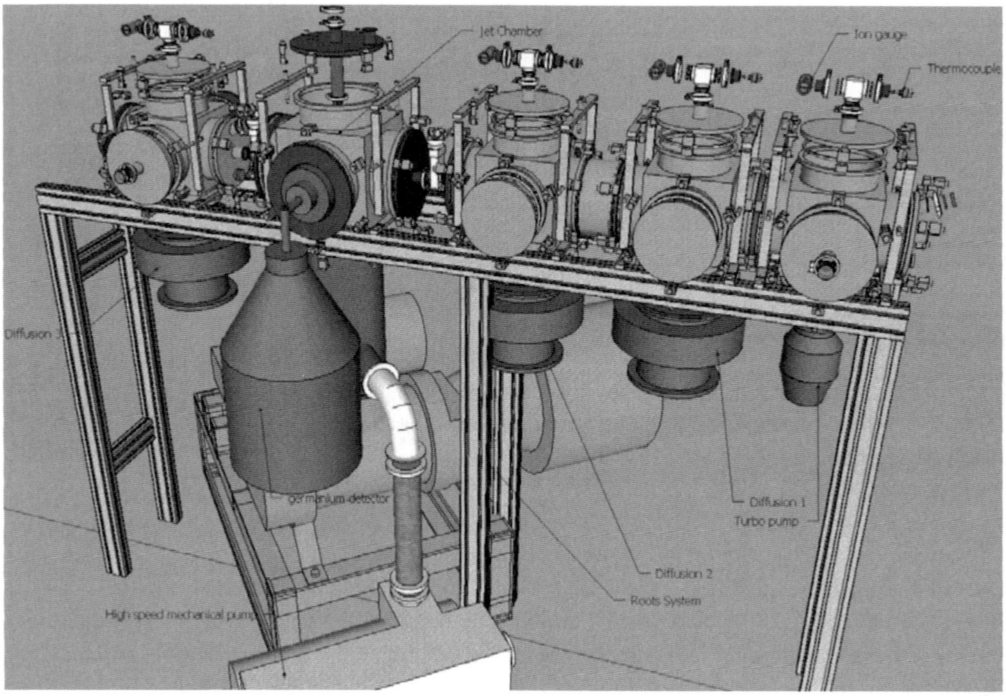

Figure 3. Differential pumping system, each chamber is separated by an inner wall with a small aperture. The jet chamber has concentric pipes, 2 stages of the differential pumping are done there.

The chambers following the scattering chamber (jet chamber), are called the differential pumping system. Each are connected with their adjacent chamber via a small aperture, big enough to allow the beam go through but small enough to limit the throughput of gas from one chamber to the neighboring ones. The first and second chamber next to the gas jet chamber are pumped down by twin diffusion pumps (1200 l/s) while the last chamber, just next to the accelerator's beam line, is pumped down by a turbo-molecular pump (400 l/s).

Using this system we are able to get the pressure in the turbo pump chamber down to the order

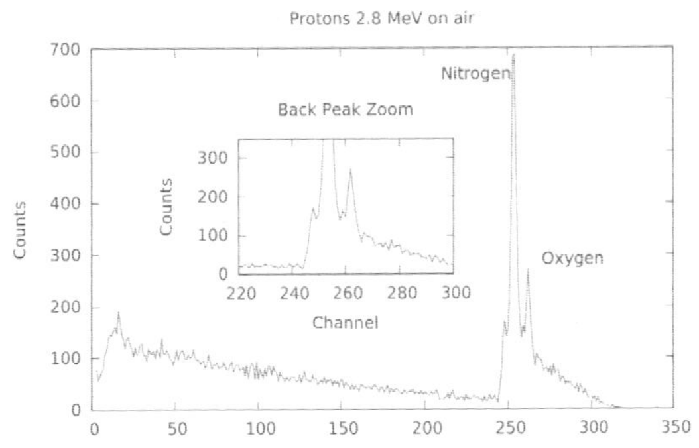

Figure 4. Protons on air and a zoom on the back peaks.

of low 10^{-6} high 10^{-7} Torr when we use air. Good references for calculating gas throughputs are [5] and [6].

Most of the pieces used in the system are ISO vacuum standard (QF and LF components), this is an important improvement from systems like this from the past (D. Shapira et.al. [1, 7], G. Bittner et. al. [8], Becker et. al. [9]) in the sense that it can be built and modified relatively quick. The used pieces are from the standard sizes LF250 (scattering chamber), LF200 (differential pumping system), LF100 (central jet pump), QF40 or QF25. And they where used from blank flanges to crosses, tees, hoses etc.

This allowed us to build the system quickly and minimized the need for special pieces. Some special pieces were needed anyway, for example, the inner walls in each chamber connect from both sides to a LF200 flange, the screws in the clamps were simply replaced with larger ones. For the apertures, a standard QF40 flange was used, and we can change the aperture size and shape simply by connecting a different QF40 flange.

Another example is in the scattering chamber, where we replace an LF250 blank flange with a special one (a nose flange) designed so that we are able to get a germanium detector as close to the jet as possible, and also allow us to place some lead shielding around the detector.

3. Results and discussion

3.1. RBS scattering

The experiments where performed with protons at various energies. And as mentioned before Air, Nitrogen and Argon where used as the gas target. In the case of Nitrogen and Argon, the back pressure is the pressure measured from the gas bottle, 1-5 atmospheres.

The RBS data from the 2 charged particle detectors was symmetrical in all of our experiments. For our first experiment air was used, we simply opened the valve of our nozzle to admit air at atmospheric pressure ($\approx 600\,Torr$).

The data is shown in figure 4. We were able to identify the various elements by their respective peaks, being the most representative the Nitrogen and the Oxygen peak.

Additionally there are a few more things to notice, one is the background. It is mostly shown in the channels before the peaks, therefore at energies below that of scattered protons.

We attribute this to the fact that there is poor vacuum in the area that surrounds the jet. So the beam actually starts scattering before it reaches the high density region of the jet. We expect that better collimation of the monitor detectors should eliminate this.

There is a small peak, more or less well defined, right behind each of the main ones (Nitrogen and Oxygen). In figure 4 we also show a zoom on the small peaks. The origin of these peaks is unclear. Since in this experiment we did not set up a set of collimators that would insure that

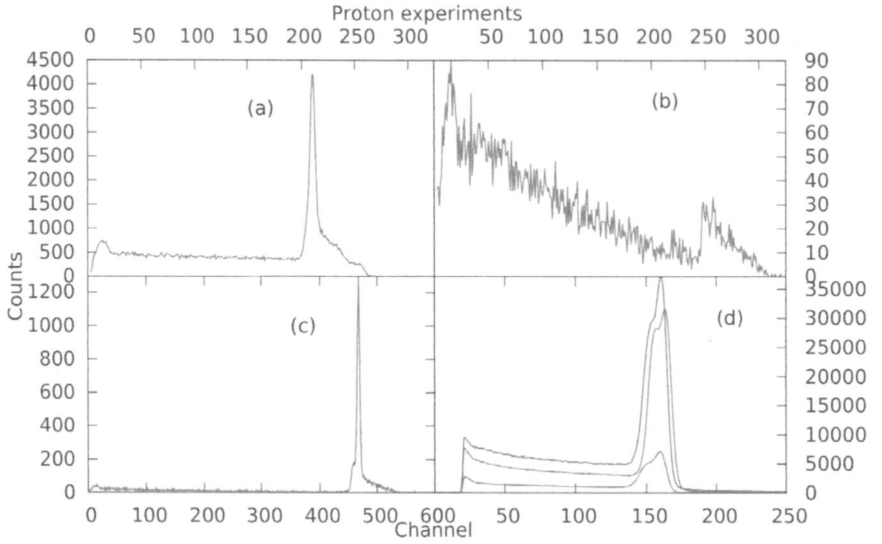

Figure 5. (a) $1.3MeV$ protons with Nitrogen target, (b) $2.8MeV$ protons without jet, (c) $2.8MeV$ protons on Nitrogen at 3.5 bars in back pressure (measured from the gas bottle), (d) count curve for $425keV$ protons at various target pressures.

particles being detected come from the beam-jet interaction region, there is room for unwanted and exotic scattering events.

Table 1 shows a comparison of the data obtained via the RBS technique *vs* the reported percentages. A software package called SIMNRA [10] was used for aid in the analysis.

Table 1. Comparison from our RBS air data *vs* reported air composition abundance.

Peak	SIMNRA	reported
Nitrogen	77%	78.09%
Oxygen	23%	20.95%

In figure 5 (a), we used $1.3MeV$ protons in Nitrogen and the peak can clearly be seen. Additionally we performed an experiment without the jet (figure 5 (b)), 2.8 MeV protons where used. The count rate was too low and most of the counts where in the low energy channels. So there is a clear difference between the jet with a high density profile and a thick gaseous target represented by the gas in the entire scattering chamber, since it is not perfect vacuum (for more info on simple gas targets see [11]).

From figure 5 (c), we can observe that:

- The main, Nitrogen, peak is well defined.
- The background, while is still non zero, it is well below the count rate of the main peak.
- A back peak can be shown in this case also (barely).

Figure 5 (d) shows the spectra of $425keV$ protons back-scattered from an Ar gas jet target produced using three different inlet pressures: 1, 2 and 3 atm. The first noticeable feature that can be seen in the figure is an energy shift or the peak as the inlet pressure is increased. This implies that the pressure in the system increases as more gas is injected in the system producing

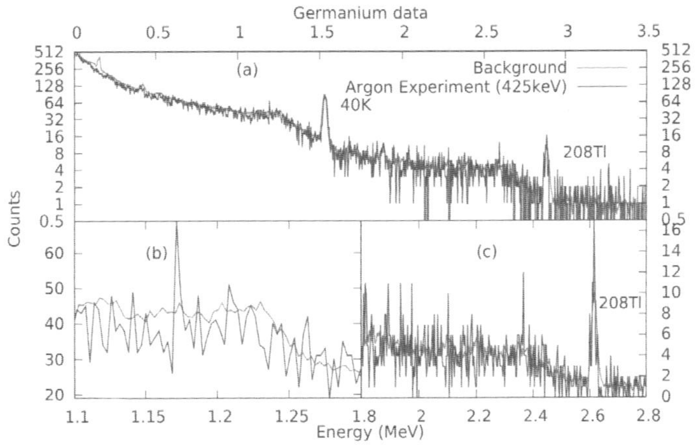

Figure 6. Germanium spectrum from 425 keV protons in Argon.

a larger energy loss in the protons. On the other hand, it can be seen that the thickness (areal density) of the target indeed increases as the inlet pressure does, as the peaks widen progressively.

3.2. Gamma experiments

To proceeded with experiments where gamma rays are detected, we removed one of the solid state detectors and replaced a side flange of the scattering chamber with a special "*nose*" flange as shown in figure 3, designed specifically to minimize the distance between the interaction region beam-gas jet target and a germanium detector.

Firstly we measured the background over 24 hours at the lab in the final configuration, that includes some lead shielding. The two main peaks where identified as Potassium 40 and Thallium 208. Then the experiment was a proton beam on an Argon gas target (3 atm) with .

The beam energy was $425keV$ as we ran as a parasite from a low energy astrophysics (p+C) experiment.

We where able to notice various peaks, mostly at energies higher that $1.5MeV$. As can be shown in figure 6 (a).

One of the regions of the spectrum where we see an excess yield, relative to the background spectrum, corresponds exactly with the ^{208}Th peak. There seems to be also a significant excess of counts in the 1.7 MeV region, as shown in the expanded regions, figure 6 (b) and (c).

It has to be said that the proton energy was much lower than the Coulomb barrier and even smaller than the Gamow energy for stars at one billion degrees temperature. More experiments are planed to be made at higher proton energies to have a clear identification of the gamma rays from the capture reaction of interest to follow them at lower energies.

4. Conclusion

A new supersonic gas jet target facility has been designed, constructed and tested at the "Universidad Nacional Autónoma de México" in the "Instituto de Física", in the "Carlos Graef Fernández" Laboratory to be used in experiments with charged particle beams provided by a 5.5 MV CN-Van de Graaff Accelerator.

In this work, the first experiments where presented showing preliminary but coherent evidence of the good performance of the system. Three different gases at several pressures were used in these tests.

It is important to notice that the RBS experiment, with a windowless pure air column, showed a close match to the expected percentage of Nitrogen and Oxygen in air.

The system is functional, the pressure readings, the picture of the jet getting hit by the beam (figure 1 (b)), the RBS data and the gamma spectrum where only a set of tests to the system.

The modular design of the system is such that it can be improved or modified easily. This gives us the freedom to change, for example, the scattering chamber without touching the differential pumping system.

These features will be of much use when we upgrade it. An important improvement to come is a recirculation system, this will allow us to work much longer with much more expensive and, in general, dangerous gases.

Additionally we intend to increase the amount of charged particle detectors in the area surrounding the jet. As well as improvements to the RBS setup, that include the use of collimators, in order to clean the spectra.

Acknowledgments

This work was supported by CONACYT 82692, PAPIIT IN-118310, the shop at IFUNAM and Dr. Roberto Gleason's support for the improvement of our facility.

References

[1] Shapira D, Jr JLCF, Novotny R, Shivakumar B, Parks RL, Thornton ST. The HHIRF supersonic gas jet target facility. Nuclear Instruments and Methods in Physics Research Section A: Accelerators, Spectrometers, Detectors and Associated Equipment. 1985;228(2–3):259 – 266. Available from: http://www.sciencedirect.com/science/article/pii/0168900285902670.

[2] Kontos A, Schürmann D, Akers C, Couder M, Görres J, Robertson D, et al. HIPPO: A supersonic helium jet gas target for nuclear astrophysics. Nuclear Instruments and Methods in Physics Research Section A: Accelerators, Spectrometers, Detectors and Associated Equipment. 2012;664(1):272 – 281. Available from: http://www.sciencedirect.com/science/article/pii/S0168900211019899.

[3] Ackroyd JAD. Fluid mechanics considerations in supersonic jet target design. Science Research Council, DARESBURY LABORATORY; 1975.

[4] Pritchard PJ. Fox and McDonald's Introduction to Fluid Mechanics. John Wiley & sons; 2011.

[5] Roth A. Vacuum technology. Second edition ed. North-Holland Physics Publishing; 1986.

[6] Lewin G. An elementary introduction to vacuum technique. American Vacuum Society monograph series. 1987;.

[7] Shapira D, Campo JGD, Jr JLCF, Shivakumar B, Stelson PH, Harmon BA, et al. Nuclear physics experiments with the ornl-hhirf supersonic gas jet target. Nuclear Instruments and Methods in Physics Research Section B: Beam Interactions with Materials and Atoms. 1985;10–11, Part 1(0):436 – 440. Available from: http://www.sciencedirect.com/science/article/pii/0168583X8590285X.

[8] Bittner G, KRETSCHMER W, SCHUSTER W. A windowless high-density gas target for nuclear scattering experiments. Nuclear Instruments and Methods. 1979;167:1–8.

[9] Becker HW, Buchmann L, Görres J, Kettner KU, Kräwinkel H, Rolfs C, et al. A supersonic jet gas target for γ-ray spectroscopy measurements. Nuclear Instruments and Methods in Physics Research. 1982;198(2–3):277–292. Available from: http://www.sciencedirect.com/science/article/pii/0167508782902654.

[10] Mayer M. SIMNRA, a simulation program for the analysis of NRA, RBS and ERDA. In: American Institute of Physics Conference Series. vol. 475 of American Institute of Physics Conference Series; 1999. p. 541–544.

[11] Baca AM. Blanco de gas sin paredes para reacciones nucleares. Tesis para obtener el título de físico, U.N.A.M. Facultad de ciencias, México D.F.; 1964.

Journal of Physics: Conference Series **492** (2014) 012011 doi:10.1088/1742-6596/492/1/012011

Data-adaptive unfolding of nuclear excitation spectra: a time-series approach

G Torres Vargas[1], R Fossion[2,3], V Velázquez[4], J C López Vieyra[5]

[1] Posgrado en Ciencias Físicas, Universidad Nacional Autónoma de México, 04510 México, D.F., Mexico

[2] Instituto Nacional de Geriatría, Periférico Sur No. 2767, 10200 México D.F., Mexico

[3] Centro de Ciencias de la Complejidad (C3), Universidad Nacional Autónoma de México, 04510 México, D.F., Mexico

[4] Facultad de Ciencias, Universidad Nacional Autónoma de México, 04510 México, D.F., Mexico

[5] Instituto de Ciencias Nucleares, Universidad Nacional Autónoma de México, 04510 México, D.F., Mexico

E-mail: fossion@nucleares.unam.mx

Abstract. A common problem in the statistical characterization of the excitation spectrum of quantum systems is the adequate separation of global system-dependent properties from the local fluctuations that are universal. In this process, called *unfolding*, the functional form to describe the global behaviour is often imposed externally on the data and can introduce arbitrarities in the statistical results. In this contribution, we show that a quantum excitation spectrum can readily be interpreted as a time series, before any previous unfolding. An advantage of the time-series approach is that specialized methods such as Singular Spectrum Analysis (SSA) can be used to perform the unfolding procedure in a data-adaptive way. We will show how SSA separates the components that describe the global properties from the components that describe the local fluctuations. The partial variances, associated with the fluctuations, follow a definite power law that distinguishes between soft and rigid excitation spectra. The data-adaptive fluctuation and trend components can be used to reconstruct customary fluctuation measures without ambiguities or artifacts introduced by an arbitrary unfolding, and also define the global level density of the excitation spectrum. The method is applied to nuclear shell-model calculations for ^{48}Ca, using a realistic force and Two-Body Random Ensemble (TBRE) interactions. We show that the statistical results are very robust against a variation in the parameters of the SSA method.

1. Introduction

According to the conjectures of Berry-Tabor [1] and Bohigas-Giannoni-Schmit [2], level fluctuations of the excitation spectra of dynamical quantum systems obey universal laws that can be described with Random Matrix Theory (RMT). These universal laws do not depend on the specific details of the particular system, and instead only depend on the underlying physical symmetries [3]; more specifically, excitation spectra of quantum systems with time-reversal symmetry are predicted to obey the statistics of the Gaussian Orthogonal Ensemble (GOE), systems without time-reversal symmetry follow the predictions of the Gaussian Unitary Ensemble (GUE), systems with time-reversal symmetry but no rotational symmetry follow the Gaussian

Journal of Physics: Conference Series **492** (2014) 012011 doi:10.1088/1742-6596/492/1/012011

Symplectic Ensemble (GSE), whereas integrable systems obey Poisson statistics (sometimes also referred to as the Gaussian Diagonal Ensemble or GDE [4]). The BGS conjecture has been confirmed in a wide variety of quantum systems, from two-dimensional systems such as quantum billiards and microwave cavities, up to many-body systems such as atoms, quantum dots, hadrons, and atomic nuclei (see [5, 6, 7, 8, 9, 10] for recent reviews). Excitation spectra of all these systems have in common that the density of states (DOS) $\rho(E)$ grows rapidly with excitation energy E because of pure combinatorial reasons, whereas the dynamics of the system (shape of mean field and/or presence of residual interactions) introduces local fluctuations around this global behaviour [11]. Thus, a statistical study of the level fluctuations of a quantum excitation spectrum offers a basic tool to elucidate the system's underlying dynamics.

Of great importance and prior to any statistical study, is the *unfolding procedure* that serves two purposes:

- to separate the system-dependent global level-density behaviour $\overline{\rho}(E)$ from the universal local fluctuations $\widetilde{\rho}(E) = \rho(E) - \overline{\rho}(E)$
- to rescale the level-spacing fluctuations to unit mean spacing so that the statistics of different systems can be compared with RMT predictions

In the traditional approach to the unfolding procedure (see e.g. [12]), a mapping of the sequence of actual energy levels $\{E(i), i = 1 \ldots N\}$, into dimensionless levels $\varepsilon(i)$,

$$E(i) \to \varepsilon(i) \equiv \overline{\mathcal{N}}\left[E(i)\right] = \int_{-\infty}^{E(i)} \overline{\rho}(E')dE', \tag{1}$$

realizes this process. Here, $\overline{\mathcal{N}}\left[E\right]$ is a smooth approximation to the accumulated or integrated density function (IDOS) $\mathcal{N}\left[E(i)\right] = \sum_{i=1}^{N} \theta\left(E - E(i)\right)$, where θ is the Heaviside step function. $\mathcal{N}\left[E(i)\right]$ counts the exact number of levels i up to excitation energy $E(i)$, and increases by one unit as the energy E passes a (non-degenerate) eigenvalue $E(i)$ [6, 7, 13]. The unfolding procedure is straightforward if a theoretical prediction for the average level density $\overline{\rho}(E)$ is available, e.g. the Weyl formula in the case of quantum billiards [9], the semi-circular distribution in the case of GOE, and the normal distribution in the Poisson case [3]. However, an analytical form for $\overline{\rho}(E)$ is usually only valid in the asymptotic limit for spectra with a very large number of levels $N \to \infty$ [8] and often is not known [7]. In those cases, a local unfolding can be applied, which can only be used to study short-range correlations [14]. Another approach, advantageous especially when the excitation spectrum contains relatively few levels, is to model the smooth behaviour of IDOS $\overline{\mathcal{N}}\left[E\right]$ rather than the global behaviour of DOS $\overline{\rho}(E)$, because the histogram representation of the latter depends on the width of the bins chosen [7]. Often, $\overline{\mathcal{N}}\left[E\right]$ is approximated by imposing a fit with a polynomial in energy E. However, the degree of the polynomial needs to be chosen carefully to avoid artifacts [10, 14]. Literature is unanimous on the fact that performing the unfolding procedure is not always an obvious task (see e.g. [9, 10, 12, 14, 15, 8, 16, 17, 18, 19]).

In [4, 10, 20], it was suggested to interpret the unfolded level fluctuations,

$$\begin{aligned} \widetilde{\mathcal{N}}\left[E(i)\right] &= \mathcal{N}\left[E(i)\right] - \overline{\mathcal{N}}[E(i)] \\ &\equiv -\delta_n = -\sum_{i=1}^{n} \left(s_i - \langle s \rangle\right) = -\left(\varepsilon_n - \varepsilon_1 - n\langle s \rangle\right) \end{aligned} \tag{2}$$

as a generalized time series. Here $s_i = \varepsilon_i - \varepsilon_{i-1}$ are the unfolded and rescaled level spacings with $\langle s \rangle = 1$. In [21], it was shown that the definitions $\widetilde{\mathcal{N}}\left[E(i)\right]$ and δ_n are equivalent. An advantage of the time-series approach is that now techniques from signal analysis from classical physics, such as Fourier spectral analysis, become available to study the quantum level fluctuations. Level fluctuations for GOE, GUE and GSE systems were found to behave as $1/f$ noise, whereas level fluctuations from systems that obey Poisson statistics behave as Brownian $1/f^2$ noise. Particular cases intermediate between Poisson and GOE behave as $1/f^\beta$ noise with $1 < \beta < 2$ [10, 20]. In [22, 23], use was made of the time-series interpretation of the unfolded level-fluctuations $\widetilde{\mathcal{N}}[E(i)]$ of eq. (2) to apply data-adaptive techniques from signal analysis to detect and correct for spurious correlations introduced by the polynomial unfolding.

The purpose of the present contribution is to show that data-adaptive techniques from signal analysis can be used not only to characterize the statistics of the unfolded level fluctuations, but to perform the very unfolding process itself. In this way, the global spectral properties and the local fluctuations are derived intrinsically, without the need to impose a user-predefined functional on the data. In fact, a similar problem in the context of time-series analysis is how to define the trend of a nonstationary time series. In a recent article [24], the importance of defining the trend in a data-adaptive way has been pointed out. Often, a requirement for the application of techniques from signal analysis is that the time series to be analyzed must be equally spaced. For this reason, we will focus on the sequence of excitation energies $\{E(i), i = 1 \dots N\}$, which is regularly sampled at steps $\Delta i = 1$, in contrast to the IDOS sequence $\mathcal{N}[E(i)]$ of eq. (1), which is not regularly sampled, $\Delta E \neq$ const, see Fig. 1. Our main point is that the sequence of actual energy levels $E(i)$ can readily be interpreted as a generalized time series, albeit with a monotonously rising behaviour. In the following, we will show that the data-adaptive technique of Singular Spectrum Analysis (SSA) [25, 26] decomposes the energy sequence $E(i)$ in a self-consistent way into a global part and a locally fluctuating part, i.e.

$$E(i) = \overline{E}(i) + \widetilde{E}(i), \tag{3}$$

which permits to extract unequivocal conclusions about the fluctuating part without the need for any previous and possibly arbitrary (local or polynomial) unfolding, as shown below.

2. Time-series analysis with Singular Spectrum Analysis (SSA)

2.1. The SSA technique

In contrast to some other data-adaptive techniques, SSA is easy to implement, can be applied to relatively short and also monotonous time series, with as only restriction that the time series needs to be regularly sampled, see Fig. 1. SSA time-series decomposition has been successfully applied in many different fields such as climatology, geology and physiology [25, 26], and the technique will be briefly discussed here. Let $f(i)$ with $i = 1, 2, \dots N$ be a discrete time series. Using the lagged vectors $X_j = (f(j), f(j+1), \dots f(j+L-1))$ with *embedding dimension* $L \leq N/2$, a so-called *trajectory matrix* $\mathbf{X} = [X_1, X_2 \dots X_K]^T$ is constructed, with $K = N - L + 1$. L is one of the few parameters of the SSA method, and we will show that in the application presented in this contribution the statistical results are very robust against a variation in the parameters. Singular Value Decomposition (SVD) is applied to $\mathbf{X}$, to identify the most meaningful basis to re-express the data. SVD decomposes $\mathbf{X}$ in a unique and exact way as a matrix product and a matrix sum $\mathbf{X} = \mathbf{U\Sigma V}^T = \sum_{k=1}^{r} \sigma_k \mathbf{X}_k$, where r is the rank of matrix $\mathbf{X}$ and $r \leq \min[K, L]$. The $K \times L$-dimensional matrix $\mathbf{\Sigma}$ contains only diagonal elements which are the ordered weights or singular values $\sigma_1 \geq \sigma_2 \geq \dots \geq \sigma_r$, and $\mathbf{X}_k = \vec{u}_k \vec{v}_k^T \equiv \vec{u}_k \otimes \vec{v}_k$ are rank-1 elementary matrices. The columns $\vec{u}_k$ of the $K \times L$-dimensional matrix $\mathbf{U}$ are the new basis vectors for the time series, and the columns $\vec{v}_k$ of the $L \times L$-dimensional matrix $\mathbf{V}$ are the associated projection coefficients. The vectors $\vec{u}_k$ can be interpreted as well as the eigenvectors of the covariance matrix $\mathbf{S}_1 = \mathbf{X} \cdot \mathbf{X}^T$ and can serve as data-adaptive filters for the time series; the associated projection coefficients can then be calculated as $\vec{v}_k = \mathbf{X}^T \cdot \vec{u}_k / \sigma_k$. Alternatively, the vectors $\vec{v}_k$ can be interpreted as the eigenvectors of the covariance matrix $S_2 = \mathbf{X}^T \cdot \mathbf{X}$, and when they are taken as data-adaptive filters then the corresponding coefficients are $\vec{u}_k = \vec{v}_k \cdot \mathbf{X}^T / \sigma_k$. The covariance matrices $\mathbf{S}_1$ and $\mathbf{S}_2$ share the same eigenvalues λ_k. The partial variance of the original time series $f(i)$ in direction $\vec{v}_k$ is given by $\lambda_k = \sigma_k^2$, whereas $\sum_{k=1}^{L} \lambda_k$ gives the total variance. Hankelization or diagonal averaging of the elementary matrix $\mathbf{X}_k$ generates the reconstructed component $g_k(i)$ and the original time series can be written exactly as the weighted sum of reconstructed components, see [25, 26],

$$f(i) = \sum_{k=1}^{L} \sigma_k g_k(i). \tag{4}$$

2.2. Trend and fluctuation analysis with SSA

SSA is a modern approach to trend extraction from a time series [27]. If the trend dominates over the fluctuations, and thus is responsible for the major part of the variance of the time series, then the trend components $g_k(i)$ are distinguished by their large partial variances λ_k [26, 27]. This is clearly the case for the energy sequence $E(i)$ or IDOS $\mathcal{N}[E]$ of a quantum excitation spectrum, which are monotonously rising functions, and often fluctuations are orders of magnitude smaller than the global behaviour, see Fig. 1 and 2. In the context of SSA, *detrending* of a time series corresponds with the separation of non-oscillating components with dominant partial variances from oscillating components with much smaller partial variances. Apart from the trend, SSA can also analyze the oscillatory components of a time series. In [19], the covariance matrices S of an ensemble of already unfolded excitation spectra corresponding to GOE and Poisson statistics were calculated. The resulting eigenvectors U_k were interpreted as *normal modes* for the level fluctuations. The diagram of ordered partial variances, or so-called *scree diagram*, was used to study the rigidity of the fluctuations. The scree diagram follows a power law,

$$\lambda_k \sim 1/k^{\gamma}, \tag{5}$$

with $\gamma \approx 1$ in the rigid case (GOE) and $\gamma \approx 2$ in the soft case (Poisson). Normal-mode analysis is related to Fourier spectral analysis: in the case of fractional Brownian motion (fBm) [28] for which the Fourier power spectrum behaves as a power law $P(f) \sim 1/f^{\beta}$, one finds a corresponding power law for the scree diagram of the form of eq. (5), with similar exponents $\beta \approx \gamma$. These results are thus in correspondence with results of refs. [4, 10, 20, 21] on the Fourier spectral analysis of time series of already unfolded quantum energy levels $\widetilde{N}[E(i)]$ (see eq. (2)) for systems obeying Poison statistics (Brownian $1/f^2$ noise) and GOE, GUE and GSE systems ($1/f$ noise).

3. Data-adaptive unfolding of quantum excitation spectra

3.1. Detrending step

When SSA is applied to a quantum excitation spectrum, then according to eq. (4), the energy sequence $\{E(i), i = 1 \ldots N\}$ can be decomposed as $E(i) = \sum_{k=1}^{L} \sigma_k g_k(i)$. The trend part,

$$\overline{E}(i) = \sum_{k=1}^{n_T} \sigma_k g_k(i), \tag{6}$$

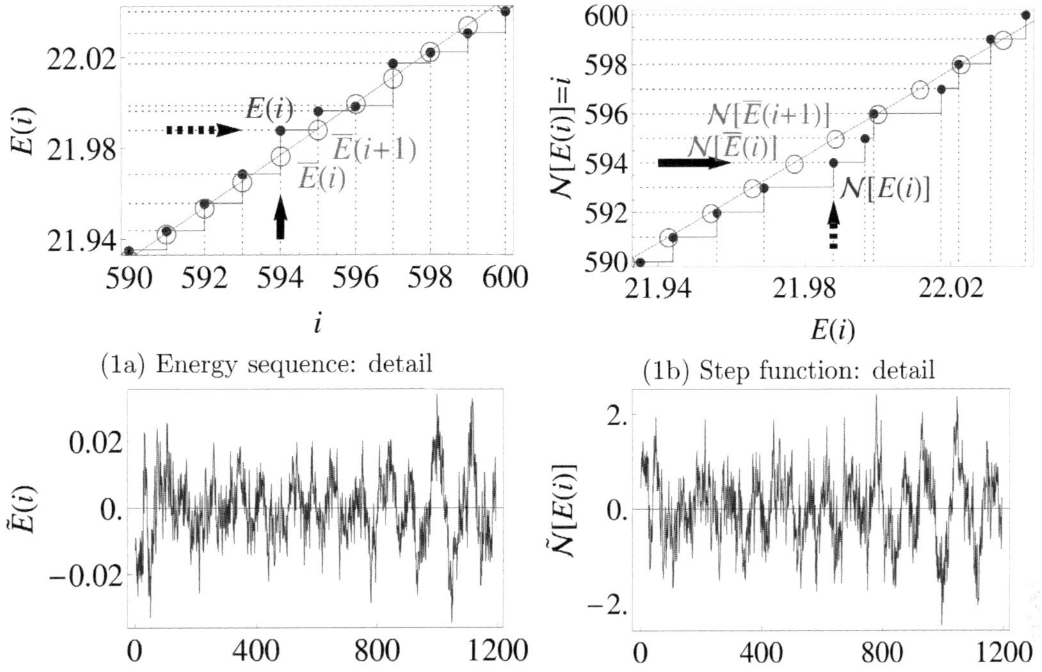

(1a) Energy sequence: detail (1b) Step function: detail

(2a) Energy sequence: fluctuations (2b) Step function: fluctuations

Figure 1. Detail of the energy sequence $E(i)$ (panel (1a)) and integrated density of states (IDOS) $\mathcal{N}[E(i)]$ (panel (1b)) for a realistic shell-model calculation of the $J^\pi = 4^+$ subspectrum of ^{48}Ca consisting of $N = 1195$ energy levels (blue filled circles). Details of the shell-model calculation are given in Sect. 4. $E(i)$ and $\mathcal{N}[E(i)]$ are compared with their respective trends (red open circles), $\overline{E}(i)$ (eq. (6)) and $\mathcal{N}[\overline{E}(i)]$ (eq. (8)), determined data-adaptively with SSA. $E(i)$ is regularly spaced, whereas $\mathcal{N}[E(i)]$ is not regularly spaced (dotted lines indicate the coordinates of the actual excitation energies). A deviation from the trend in the horizontal direction is necessarily accompanied with a deviation in the vertical direction, as indicated by the arrows for a particular fluctuation at $i = 594$. Thus, apart from obvious differences in sign and scale, energy-sequence fluctuations $\widetilde{E}(i) = E(i) - \overline{E}(i)$ (panel (2a) and eqs. (3) and (7)) and IDOS fluctuations $\widetilde{\mathcal{N}}[E(i)] = \mathcal{N}[E(i)] - \mathcal{N}^{\text{intrpl}}_{E(i)}[\overline{E}(i)]$ (panel (2b) and eq. (9)) are statistically equivalent. Note that the fluctuations $\widetilde{E}(i)$ and $\widetilde{\mathcal{N}}[E(i)]$ are orders of magnitude smaller than the global behaviour of $E(i)$ and $\mathcal{N}[E(i)]$, respectively.

is separated from the fluctuation part,

$$\widetilde{E}(i) = \sum_{k=n_T+1}^{L} \sigma_k g_k(i) \tag{7}$$

and the number of components n_T to be included into the trend is defined in a data-adaptive way, by identifying the dominant partial variances $\lambda_k = \sigma_k^2$ with $k = 1 \ldots n_T$, see Fig. 3. This detrending corresponds to the first step of the unfolding procedure. Already during the detrending step, the power law $\lambda_k \sim 1/k^\gamma$ of eq. (5) of the fluctuation part of the scree diagram distinguishes between soft ($\gamma \approx 2$) and rigid ($\gamma \approx 1$) spectra, see Fig. 3. If desired, the trend part $\overline{E}(i)$ can be used to calculate the global DOS $\overline{\rho}(E) = \rho(\overline{E})$ of the quantum excitation spectrum, see Fig. 2. On the other hand, specialized techniques from signal analysis can be applied to the fluctuation part $\widetilde{E}(i)$ to further characterize the fluctuation statistics. An advantage of using time-series based fluctuation measures, such as normal-mode analysis [19], Fourier spectral analysis [4, 18] or Detrended Fluctuation Analysis (DFA) [18, 29], is that $\widetilde{E}(i)$ does not need to be rescaled prior to analysis, because in the case of power-law behaviour the scale is absorbed in the offset, whereas information on the correlations is codified in the exponent of the power law, see Fig. 3 and 4 (panels (1)). In principle, if one could unfold both the step function $\mathcal{N}[E(i)]$ and the energy sequence $E(i)$ of a given excitation spectrum without ambiguities, the fluctuations obtained, $\widetilde{N}[E(i)]$ and $\widetilde{E}(i)$, respectively, would be statistically similar, see Fig. 1. The most important difference being that the former fluctuations are detrended and rescaled, whereas the latter fluctuations are only detrended.

3.2. Rescaling step

Some of the traditional spectral fluctuation measures require an explicit prior rescaling of the level fluctuations, as is the case of the short-range correlation measure of the Nearest Neighbour Spacing (NNS) distribution, and the long-range correlation measure of the Δ_3 statistic [3]. Such rescaling is traditionally achieved by projecting the excitation energies over the smooth trend $\overline{\mathcal{N}}[E(i)]$ fitted to the integrated density function (IDOS) $\mathcal{N}[E(i)]$, see eq. (1). In the present case, this smooth trend is instead given data-adaptively by integrating the SSA energy trend $\overline{E}(i)$, i.e. by the step function $\mathcal{N}[\overline{E}]$. For the projection, we then get,

$$E(i) \to \eta(i) = \mathcal{N}^{\text{intrpl}}_{E(i)}[\overline{E}(i)] = \mathcal{N}[\overline{E}(i)] + \frac{E(i) - \overline{E}(i)}{\overline{E}(i+1) - \overline{E}(i)}, \tag{8}$$

where a linear interpolation has been used in between trend values $\mathcal{N}[\overline{E}(i)]$ and $\mathcal{N}[\overline{E}(i+1)]$, see Fig. 1. Following the mapping of eq. (8), the detrended and rescaled level fluctuations are given by,

$$\widetilde{\mathcal{N}}[E(i)] = \mathcal{N}[E(i)] - \mathcal{N}^{\text{intrpl}}_{E(i)}[\overline{E}(i)], \tag{9}$$

from which the Δ_3 statistic can be calculated, see Fig. 4 (panels (2)). The corresponding level spacings are given by,

$$s(i) = \eta(i) - \eta(i-1). \tag{10}$$

By construction, the level spacings $s(i)$ are now normalized to unit mean spacing $\langle s \rangle = 1$, as required by the NNS statistic, see Fig. 4 (panels (3)).

4. Results: Application to the excitation spectrum of the atomic nucleus ^{48}Ca

The present unfolding method, based on the SSA time-series data-adaptive technique, is tested for the excitation spectrum of the ^{48}Ca atomic nucleus. Fluctuation studies must be applied to levels with the same quantum numbers; we choose the $J^{\pi} = 4^{+}$ subspectrum because it has the largest number of levels. The level fluctuations of ^{48}Ca have been thoroughly studied in literature and have been found to correspond with GOE predictions [10, 18, 22, 30, 31], which permits to verify the validity of the SSA unfolding method. The energy sequence $\{E(i), i = 1 \ldots N\}$ is calculated with a nuclear shell-model code, based on ANTOINE [32]. ^{48}Ca is an obvious choice to study within the nuclear-shell model framework, because an inert ^{40}Ca core can be assumed, and calculations can be carried out in the full neutron fp shell. All figures in this contribution are

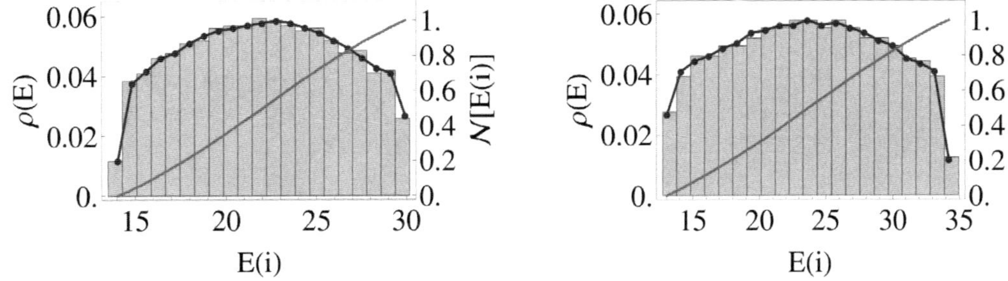

(a) One realization with a realistic calculation (b) One realization of the TBRE ensemble

Figure 2. Density of states (DOS) $\rho(E) = \mathrm{d}\mathcal{N}(E)/\mathrm{d}E = \sum_i \delta(E - E(i))$ [13] (histogram), compared with the mean DOS $\overline{\rho}(E) = \rho(\overline{E}) = \mathrm{d}\mathcal{N}(\overline{E})/\mathrm{d}E = \sum_i (E - \overline{E}(i))$ calculated data-adaptively with SSA from the trend $\overline{E}(i)$ of eq. (6) (blue line), for the $J^\pi = 4^+$ subspectrum of ^{48}Ca. The integrated density of states (IDOS) $\mathcal{N}(E) = \overline{\mathcal{N}}(E) + \widetilde{\mathcal{N}}(E)$ (red line) is a monotonously rising function, with a dominant global trend $\overline{\mathcal{N}}(E)$. For representational purposes, in these figures, $\mathcal{N}(E)$ has been normalized such that $\mathcal{N}(E_{\max}) = 1$.

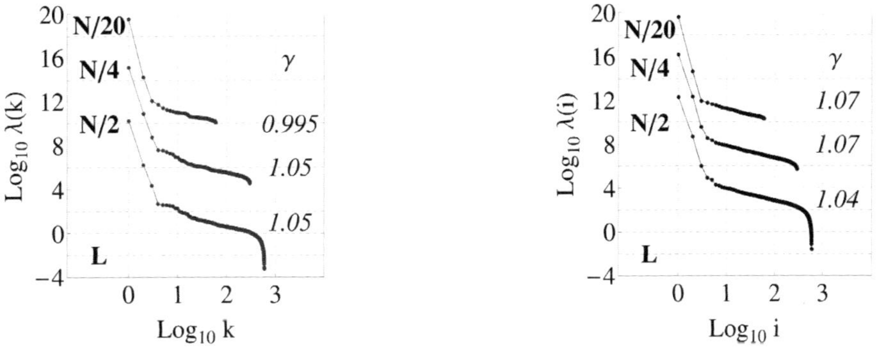

(a) Realistic calculation (b) TBRE ensemble

Figure 3. Scree diagram of ordered partial variances λ_k resulting from the SSA analysis of the energy sequence $\{E(i), i = 1 \dots N\}$ for the $J^\pi = 4^+$ subspectrum of ^{48}Ca, in case of (a) a single realization of the spectrum using a realistic interaction (blue), and (b) an ensemble mean with $n = 8$ realizations of the spectrum using Two-Body Random Ensemble (TBRE) interactions (black). Details of the calculation are presented in Sect. 4. The largest λ_k with $k = 1, \dots n_T$ are separated by orders of magnitude from the smaller λ_k and correspond to the trend $\overline{E}(i)$, see eq. (6). The smaller λ_k with $k = (n_T + 1), \dots L$ correspond to the fluctuations $\widetilde{E}(i)$, see eq. (7), and follow a power law $\lambda(k) \sim 1/k^\gamma$ with exponent $\gamma \approx 1$, see eq. (5), indicating a rigid quantum excitation spectrum. The power law is independent from the embedding dimension L. Shown for $L = N/2, N/4$ ($n_T = 3$) and $N/20$ ($n_T = 2$). The different scree diagrams have been shifted vertically for reasons of comparison.

based on data from this shell-model calculation, for a single realization of the spectrum using a realistic interaction (KB3 [33, 34]) (left panels), and for an ensemble of $n = 8$ realizations using random interactions (TBRE [21, 35]) (right panels). To take into account only the bulk (central part) of the spectrum, 16% of the lower and upper levels have been discarded, and the unfolding is applied to a total of $N = 1195$ levels.

In Fig. 3, the scree diagram of ordered partial variances λ_k is shown for three different choices of

Journal of Physics: Conference Series **492** (2014) 012011 doi:10.1088/1742-6596/492/1/012011

embedding dimension L. It can be seen that in all three cases, the first $n_T = 2$ (for $L = N/20$) or $n_T = 3$ (for $L = N/4, N/2$) largest partial variances λ_k separate from the rest of the eigenvalues (note the double logarithmic scale!). These λ_k with $k = 1, \ldots, n_T$ correspond with the reconstructed components $g_k(i)$ that carry most ($> 99.99\%$) of the variance of the energy sequence $E(i)$, and constitute its monotonous and non-oscillating trend $\overline{E}(i)$, see eq. (6). The smaller eigenvalues λ_k with $k = (n_T + 1), \ldots L$ behave as the power law of eq. (5) with exponent $\gamma \approx 1$, indicating a rigid power spectrum, as expected for the $J^\pi = 4^+$ subspectrum of ^{48}Ca [18]. It is easy to check that the corresponding reconstructed components $g_k(i)$ oscillate with an average frequency proportional to k, and they describe the local level fluctuations $\widetilde{E}(i)$ of eq. (7). It is worth noticing that the power-law exponent γ does not depend on the particular choice of embedding dimension L, and does not depend on the number of trend components n_T either, as long as $n_T \geq 2$ or $n_T \geq 3$, respectively. For very large L ($L \approx N/2$), the decomposition includes a large number of insignificantly small eigenvalues, whereas for very small L ($L \approx N/20$), the range over which the power-law behaviour is observed is reduced. In the present contribution, we use an intermediate choice, namely $L = N/4$ (also recommended in [25]).

In Fig. 2, the DOS $\rho(E)$ is shown for the $J^\pi = 4^+$ subspectrum of ^{48}Ca, in comparison with the global DOS $\overline{\rho}(E)$ as follows from the SSA data-adaptive trend $\overline{E}(i)$ from eq. (6). In the case of atomic nuclei, understanding the level density is very important in order to understand compound nuclear reactions. Two important examples of nuclear reactions are the decay of the giant-dipole resonance in hot nuclei, and the radiative capture of protons, neutrons, $\alpha-$particles, etc., in nucleosynthesis. In order to have predictions in nuclei where no experimental information is available, the global level density $\overline{\rho}(E)$ can be modelled by theory, using e.g. the Bethe formula [36], the binomial distribution [37], or Monte-Carlo shell-model calculations [38]. The SSA defined global level density $\overline{\rho}(E)$ permits to check the adequacy of the predicted level densities for some carefully chosen exerimental test cases where $\rho(E)$ is known.

In Fig. 4 (panels (1)), the Fourier power spectrum of the detrended fluctuation series $\widetilde{E}(i)$ of eq. (7) is presented. Results are shown for $L = N/4$ and for $n_T = 3, 5$ and 10. In all cases, the power spectrum corresponds with $1/f$ noise ($\beta \approx \gamma \approx 1$), indicating a rigid spectrum, independently from the specific choice of the number of trend components n_T. Only the range over which the power law is observed, diminishes when more monotonous or slowly oscillating components are included in the trend (excluded from the fluctuations).

In Fig. 4 (panels (2)), the long-range fluctuations statistic Δ_3 is shown for the detrended and rescaled level fluctuations $\eta(i)$ of eq. (8), and is found to correspond well with the GOE prediction. Results are shown for $L = N/4$ and $n_T = 3, 5$ and 10, and results are found to be very robust against the specific choice of the parameters, with as only difference that the large-scale behaviour of Δ_3 flattens out when more monotonous or slowly oscillating components are included in the trend (excluded from the fluctuations).

In Fig. 4 (panels (3)), the short-range fluctuation statistic of the NNS distribution $P(s)$ is shown for the detrended and rescaled the level spacings $s(i) = \eta(i) - \eta(i-1)$ of eq. (10), and corresponds well with the Wigner surmise $P_W(s) = (\pi s/2)\exp(-\pi s^2/4)$ for GOE. Results are shown for $L = N/4$ and $n_T = 3$, but results are found to be independent from the specific choice of the parameters.

5. Conclusions

We have proposed a method to realize a data-adaptive unfolding of quantum excitation spectra, which avoids the possible ambiguities when the unfolding functional form is imposed externally to the data (e.g. using a polynomial of arbitrary degree). We interpreted the sequence of actual energy levels $\{E(i), i = 1 \ldots N\}$ as a generalized time series, and used the data-adaptive SSA method to separate in an unambiguous way the global trend $\overline{E}(i)$ from the local fluctuations $\widetilde{E}(i)$. This first step corresponds with the detrending part of the unfolding procedure. The statistical results are very robust against a variation in the two parameters of the method (embedding dimension L and number of trend components n_T). The partial variances, associated to the fluctuations, follow a power law that distinguishes between soft and rigid excitation spectra. If desired, the data-adaptive trend components $\overline{E}(i)$ can be used to calculate the global level density behaviour, and specialized techniques from signal analysis can be applied to the data-adaptive fluctuation components $\widetilde{E}(i)$ to further specify the fluctuation statistics. Also the rescaling step of the unfolding procedure can be carried out in a self-consistent way, after which traditional fluctuation measures such as the Nearest-Neighbour Spacing (NNS) distribution and Δ_3 can be calculated without ambiguities. We tested the present data-adaptive unfolding method for the $J^\pi = 4^+$ excitation subspectrum of the ^{48}Ca atomic nucleus, and our results are in correspondence with previous results in literature. This also illustrates the fact that the fluctuations in the excitation energies $\widetilde{E}(i)$ (considered in the present contribution) are statistically equivalent to the fluctuations $\widetilde{\mathcal{N}}[E(i)]$ of the integrated density function (IDOS) (considered traditionally). In a forthcoming publication, we will show that the proposed unfolding method is general enough to apply to the excitation spectra of other quantum systems, and also to the eigenspectra of random matrices in various applications in quantum and classical physics [39].

We acknowledge financial support from CONACYT (grants CB-2011-01-167441, CB-2010-01-155663, I010/266/2011/C-410-11 and grant 351103 from the Red Temática Envejecimiento, Salud y Desarrollo Social), PAPIIT-DGAPA (grant IN114411), the European Commission (project FP7-PEOPLE-2009-IRSES-247541-MATSIQEL) and the Instituto Nacional de Geriatría (project DI-PI-002/2012). The authors wish to thank Dr. A. Frank and collaborators for fruitful discussions.

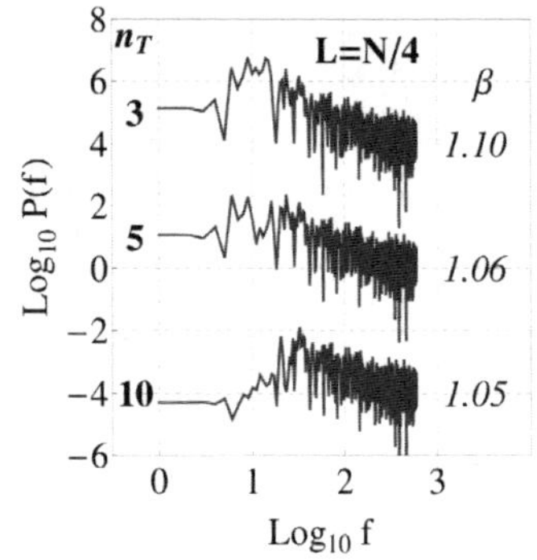

(1a) Realistic calculation: power spectrum

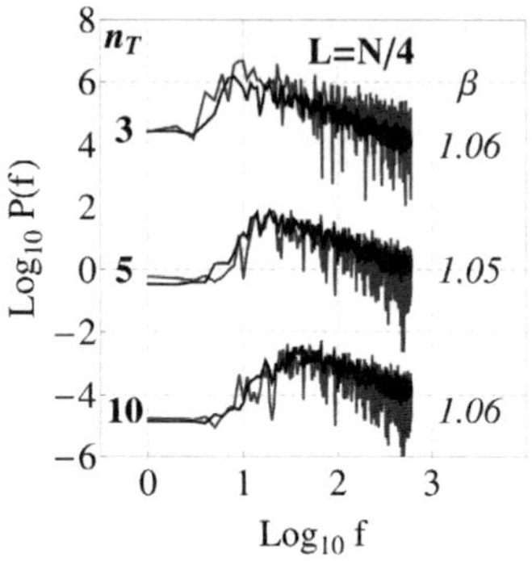

(1b) TBRE ensemble: power spectrum

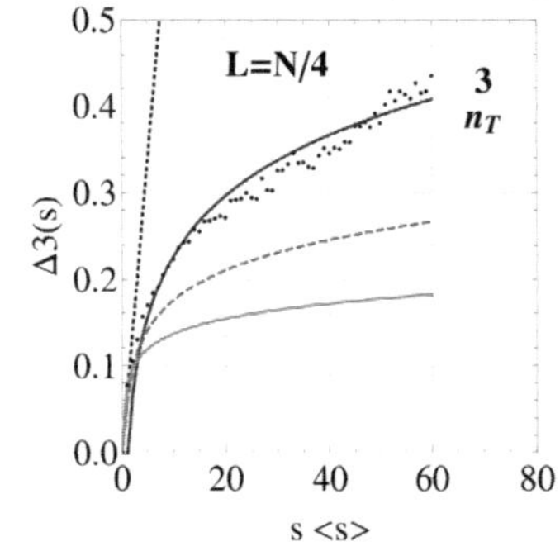

(2a) Realistic calculation: Δ_3

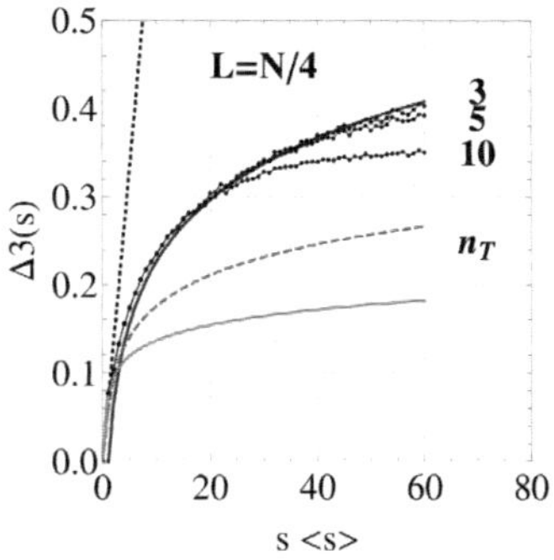

(2b) TBRE ensemble: Δ_3

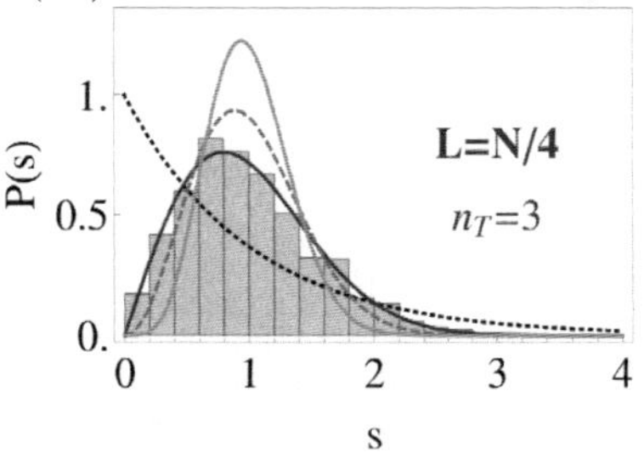

(3a) Realistic calculation: NNS

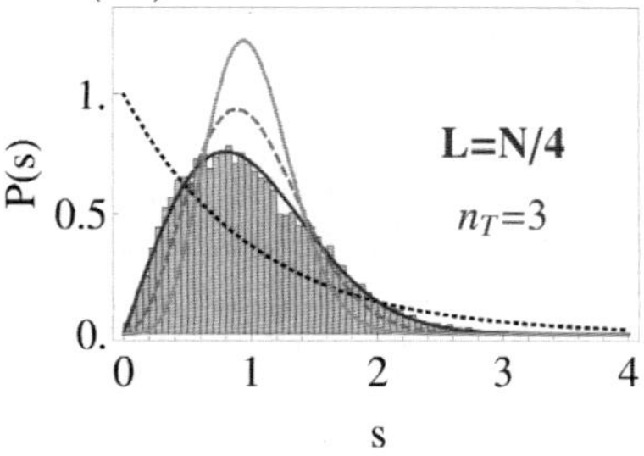

(3b) TBRE ensemble: NNS

Figure 4. Fluctuation measures for the theoretical $J^\pi = 4^+$ subspectrum of ^{48}Ca, for a single realization using a realistic interaction (left panels (a)) and a Two-Body Random Ensemble (TBRE) (right panels (b)). (Upper row (1)) The Fourier power spectrum $P(f)$ for the detrended fluctuation series $\widetilde{E}(i)$ of eq. (7) follows a power law $P(f) \sim 1/f^\beta$ with $\beta \approx \gamma \approx 1$. Shown for embedding dimension $L = N/4$ and for number of trend components $n_T \geq 3$. The power law is independent from the particular choice of the parameters L and n_T. Shown for a single realization of the spectrum (blue), and for an ensemble mean (black). (Middle panels (2)) The long-range fluctuation measure Δ_3 for the detrended and rescaled energies $\eta(i)$ of eq. (8) corresponds with the GOE prediction (blue, continuous line), compared with predictions for the Poisson (black, dotted line), GUE (red, dashed line) and GSE (orange, continuous line) cases. Shown for embedding dimension $L = N/4$ and number of trend components $n_T = 3, 5$ and 10. (Bottom panels (3)) The short-range fluctuation measure of the Nearest-Neighbour Spacing (NNS) distribution for the detrended and rescaled level spacings $s(i) = \eta(i) - \eta(i-1)$ of eq. (10) corresponds with the GOE prediction (blue, continuous line), compared with predictions for the Poisson (black, dotted line), GUE (red, dashed line) and GSE (orange, continuous line) cases. Shown for embedding dimension $L = N/4$ and number of trend components $n_T = 3$. The results are independent from the embedding dimension L and the number of trend components n_T. It can be appreciated that the level spacings s have been normalized to $\langle s \rangle = 1$.

References

[1] Berry M V and Tabor M 1977 *Proc. Roy. Soc. A* **356** 375
[2] Bohigas O, Giannoni M J and Schmit C 1984 *Phys. Rev. Lett.* **52** 1
[3] Mehta M L 1991 *Random Matrices* 2nd ed. (New York: Acad. Press)
[4] Relaño A, Gómez J M G, Molina R A, Retamosa J and Faleiro E 2002 *Phys. Rev. Lett.* **89** 244102
[5] Kota V B K 2001 *Phys. Rep.* **347** 223
[6] Papenbrock T and Weidenmüller H A 2007 *Rev. Mod. Phys.* **79** 997
[7] Weidenmüller H A and Mitchell 2009 G E *Rev. Mod. Phys.* **81** 539
[8] Guhr T, Müller-Groeling A and Weidenmüller H A 1998 *Phys. Rep.* **299** 189
[9] Haake F 2010 *Quantum signatures of chaos* 3rd ed. (Heidelberg: Springer)
[10] Gómez J M G, Kar K, Kota V B K, Molina R A, Relaño A and Retamosa J 2011 *Phys. Rep.* **499** 103
[11] Zelevinsky V 2001 *Phys. E* **9** 450.
[12] Brody T A, Flores J, French J B et al. 1981 *Rev. Mod. Phys.* **53** 385
[13] Shukla P 2012 *Int. J. Mod. Phys. B* **26** 1230008
[14] Gómez J M G, Molina R A, Relaño A and Retamosa J 2002 *Phys. Rev. E* **66** 036209
[15] Bae M S, Otsuka T, Mizusaki T and Fukunishi N 1992 *Phys. Rev. Lett.* **69** 2349
[16] Enders J, Guhr T, Heine A et al. 2004 *Nucl. Phys. A* **741** 3
[17] Fossion R and Bijker R 2009 *Rev. Mex. Fís.* S **55** 41
[18] Landa E, Morales I O, Hernández C, López Vieyra J C and Frank A 2008 *Rev. Mex. Fís.* S **54** 48
[19] Jackson A D, Mejia-Monasterio C, Rupp T, Saltzer M and Wilke T2001 *Nucl. Phys. A* **687** 405
[20] Gómez J M G, Relaño A, Faleiro E et al. 2005 *Phys. Rev. Lett.* **94** 084101
[21] Relaño A, Molina R A and Retamosa J 2004 *Phys. Rev. E* **70** 017201
[22] Landa E, Morales I O, Fossion R et al. 2011 *Phys. Rev. E* **84** 016224
[23] Morales I O, Landa E, Stránský P and Frank A 2011 *Phys. Rev. E* **84** 016203
[24] Wu Z, Huang N E, Long S R and Peng Ch-K 2007 *Proc. Nat. Acad. Sci.* **104** 14889
[25] Elsner J B and Tsonis A A 1996 *Singular Spectrum Analysis:* (New York: Plenum Press,)
[26] Golyandina N et al. 2001, *Analysis of time series structure* (London: Chapman & Hall/CRC).
[27] Alexandrov T, Bianconcini S, Dagum E B, Maass P and McElroy T 2012 *Econometric Rev.* **31** 593
[28] Gao J B, Cao Y and Lee J M 2003 *Phys. Lett. A* **314** 392
[29] Santhanam M S, Bandyopadhyay J N and Angom D 2006 *Phys. Rev. E* **73** 015201(R)
[30] Gómez J M G et al. 1998 *Phys. Rev. C* **58** 2108.
[31] Caurier E , J.M.G. Gómez, V.R. Manfredi and L. Salasnich et al. 1996 *Phys. Lett. B* **365** 7
[32] Caurier E and Nowacki F 1989 code Antoine, Strasbourg
[33] Kuo T et al. 1968 *Nucl. Phys. A* **114** 241.
[34] Poves A et al. 1981 *Phys. Rep.* **70** 235
[35] Flores J, Horoi M, Müller M and Seligman T H 2001 *Phys Rev E* **63** 026204
[36] Bethe H A 1936 *Phys. Rev.* **50** 332
[37] Zuker A P 2001 *Phys. Rev. C* **64** 021303
[38] Dean D J, Koonin S E, Langanke K, Radha P B and Alhassid Y 1995 *Phys. Rev. Lett.* **74** 2909
[39] Fossion R, Torres Vargas G, Velázquez V and López Vieyra J C, in preparation

Journal of Physics: Conference Series **492** (2014) 012012 doi:10.1088/1742-6596/492/1/012012

Fidelity, susceptibility and critical exponents in the Dicke model

M A Bastarrachea-Magnani, O Castaños, E Nahmad-Achar, R López-Peña and J G Hirsch

Instituto de Ciencias Nucleares, Universidad Nacional Autónoma de México
Apdo. Postal 70-543, México D. F., C.P. 04510

E-mail: miguel.bastarrachea@nucleares.unam.mx

Abstract. We calculate numerically the fidelity and its susceptibility for the ground state of the Dicke model. A minimum in the fidelity identifies the critical value of the interaction where a quantum phase crossover, the precursor of a phase transition for finite number of atoms $\mathcal{N}$, takes place. The evolution of these observables is studied as a function of $\mathcal{N}$, and their critical exponents evaluated. Using the critical exponents the universal curve for the specific susceptibility is recovered. An estimate to the precision to which the ground state wave function is numerically calculated is given, and found to have its lowest value, for a fixed truncation, in a vicinity of the critical coupling.

1. Introduction

The Dicke Hamltonian describes a system of $\mathcal{N}$ two-level atoms interacting with a single monochromatic electromagnetic radiation mode within a cavity [1]. In terms of quantum computation, it can also describe a set of $\mathcal{N}$ qubits, realized with quantum dots, Bose-Einstein condensates or QED circuits [2], interacting through a bosonic field. In recent years Dicke-like Hamiltonians, and in particular its quantum phase transition (QPT) from normal to superradiant behavior [3, 4], have attracted much attention. The QPT is an example of a quantum collective behavior and has a close connection with entanglement and quantum chaos. Besides, the Dicke Hamiltonian for a finite $\mathcal{N}$ provides a good description for the systems manipulated in the laboratory, especially in the light of the experimental realization of the superradiant phase transition in a BEC [7, 8], the intense development in the control of single atoms and photons in a cavity, and the possibility of a QPT in a system of $\mathcal{N}$ QED circuits [5, 6].

The Dicke model can be written as ($\hbar = 1$)

$$H_D = \omega a^\dagger a + \omega_0 J_z + \frac{\gamma}{\sqrt{\mathcal{N}}}\left(a + a^\dagger\right)(J_+ + J_-), \quad (1)$$

where ω is the field frequency, ω_0 is the atomic energy separation, $a^\dagger$ and a are the creation and annihilation photon operators, respectively, and γ is the coupling strength. J_z and $J_\pm$ are collective atomic operators (pseudospin operators) which follow the SU(2) algebra, and denote the atomic relative population and the atomic transitions operators, respectively.

For a finite number of atoms $\mathcal{N}$, the model is in general non-integrable, and care must be taken when the first order in the $1/\mathcal{N}$ expansion is employed because of its singular behavior around the phase transition [9, 10]. The Hamiltonian is integrable in at least two limits ($\gamma \to 0$ and $\omega_0 \to 0$). In the thermodynamic limit, when the number of atoms $\mathcal{N}$ goes to infinity, the mean field description becomes exact. It provides analytic asymptotic solutions through a Holstein-Primakoff expansion [11], which allows to extract the critical exponents for the ground state energy per particle, the fraction of excited atoms, the number of photons per atom, their fluctuations and the concurrence [11, 12, 13, 14]. Numerical solutions complement and confirm the theoretical predictions, and allow for the exploration of the system in regimes which are not described by the latter, like excited state phase transitions.

A concept emerging from quantum information theory, the fidelity, can be used to determine a sudden change in the ground state of a quantum system as a function of a control parameter. In recent years it has emerged as a powerful tool to study QPT in quantum many-body systems [15]. The fidelity describes the overlap between two quantum states. Considering a quantum many-body system, the general form of the Hamiltonian can be written as

$$H = H_0 + \gamma H_1, \tag{2}$$

where H_1 is the interaction Hamiltonian and γ is a control parameter. For two pure states $|\psi(\gamma)\rangle$ and $|\psi(\gamma')\rangle$ the fidelity is written as [15]

$$F(\gamma,\gamma') = |\, \langle\psi(\gamma)\, |\psi(\gamma')\rangle\, |. \tag{3}$$

The fidelity measures the amount of shared information between two quantum states, being its geometric interpretation the closeness of these states. Being a QPT a sudden change in the ground state properties of a system when a control parameter varies, a minimum in the fidelity allows to locate and characterize the QPT. Its second derivative, the fidelity susceptibility, is even more sensitive to the QPT. Expanding the fidelity around its minimum, for $\gamma - \gamma'$ small, we have [15]

$$F(\gamma,\gamma') = 1 - \frac{(\gamma-\gamma')^2}{2}\chi^F + ... \tag{4}$$

The fidelity susceptibility χ^F can be expressed as

$$\chi^F(\gamma) = \lim_{\gamma-\gamma'\to 0} \frac{-2\ln F(\gamma,\gamma')}{(\gamma-\gamma')^2} = \frac{2(1-F(\gamma,\gamma'))}{(\gamma-\gamma')^2}, \tag{5}$$

being the first form in terms of the logarithmic fidelity. It is useful to choose $\gamma' = \gamma + d\gamma$ in order to vary γ while taking the limit $d\gamma \to 0$.

In the thermodynamic limit, the fidelity goes to zero in the QPT, while the susceptibility goes to infinity. For finite systems, in the critical value of the coupling γ_{max}, the fidelity and its susceptibility show the *precursor* of the QPT by obtaining a minimum and a maximum value, respectively. Calculating the behavior of these quantities (the critical coupling parameter and the maximum value of the susceptibility) allows us to derive their critical exponents as a function of the number of atoms $\mathcal{N}$ [16, 17]. Furthermore, one can obtain universal curves for some observables like the fidelity [18] or the susceptibility. For a finite scale analysis, we can define a universal quantity called the specific susceptibility [19],

$$\chi_s = \frac{\chi^F(\gamma_{max}) - \chi^F(\gamma)}{\chi^F(\gamma)}. \tag{6}$$

The specific susceptibility is useful to compare systems with different number of atoms.

In this work we calculate the fidelity and its susceptibility for the ground state of the finite Dicke model, performing a numerical diagonalization of the Hamiltonian. Using the fidelity formalism we locate the precursor of the QPT for each $\mathcal{N}$. With it, we find numerically the critical exponent of the coupling parameter, which tends to $\gamma_c = \sqrt{\omega\omega_0}/2$, the critical value in the thermodynamic limit. We also study the behavior of the minimum of the fidelity and the maximum of its susceptibility as $\mathcal{N}$ grows, finding their critical exponents. We build the universal curve of the specific susceptibility, which confirms the value of the critical exponent. Finally, we make a brief discussion of the ground state wave function precision as a function of the coupling strength.

2. Numerical solution

In order to solve numerically the Dicke Hamiltonian we employ extended bosonic coherent states [14, 20]. They are built with the displaced boson operators $A^\dagger, A$, obtained by shifting the original annihilation operator a:

$$A = a + \frac{2\gamma}{\omega\sqrt{\mathcal{N}}} J_x. \tag{7}$$

The new basis is $\{|N; j, m\rangle\}$, where N is an eigenvalue of the new number operator $A^\dagger A$, $j = \mathcal{N}/2$ and m is an eigenvalue of J_x. It allows for the determination of ground state properties in the superradiant region far beyond previous attempts [14], and also of excited states with a single truncation [21].

To solve the Hamiltonian numerically we must truncate the Hilbert space, which is infinite due to the presence of the number operator in the Hamiltonian. In order to estimate the minimal truncation to be employed, we use a criterion based on the precision of the wave function, which we call the ΔP criterion [22]. We express the ground state wave function as:

$$|\Psi(N_{max})\rangle = \sum_{N=0}^{N_{max}} \sum_{m=-j}^{j} C_{N,m} |N; j, m\rangle, \tag{8}$$

where $C_{N,m}$ are the coefficients of the exact ground state wave function and N_{max} is the value of the truncation in the number of displaced excitations. The probability P_N of having N excitations in the ground state is:

$$P_N = |\langle N|\Psi\rangle|^2 = \sum_m |C_{N,m}|^2 \tag{9}$$

We define the precision in the calculated wave function as (see Appendix)

$$\Delta P = \sum_{m=-j}^{j} |C_{N_{max}+1,m}(N_{max}+1)|^2 . \tag{10}$$

By diagonalizing the Hamiltonian with several truncations, if ΔP is smaller than certain tolerance we consider that the solution has converged, being N_{max} the minimum value of the truncation necessary for obtaining the exact numerical solution.

3. Results

We calculate the fidelity and its susceptibility as functions of the coupling γ for the ground state by solving numerically the Hamiltonian. In figures 1 and 2 we show the fidelity for several values of $\mathcal{N}$ from 100 to 1000. The same goes for the fidelity susceptibility in figures 3 and 4. In these

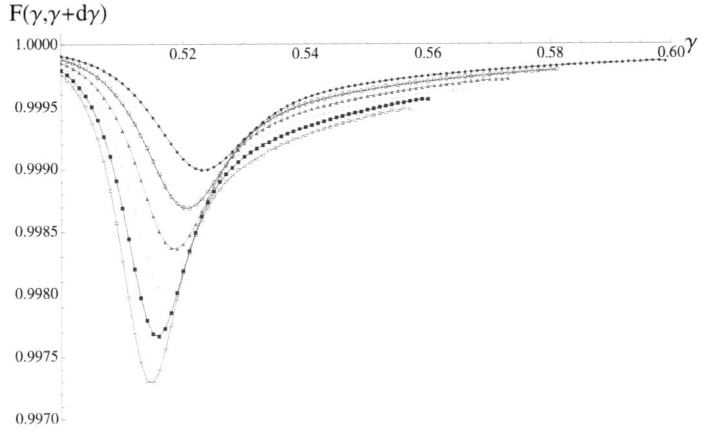

Figure 1. Fidelity as a function of the coupling parameter. With γ from $\gamma_c = 0.5$ to 0.6, $d\gamma = 0.001$, $\omega_0 = \omega = 1$ and $\mathcal{N} = 100$, 120, 140, 160, 180, 200 (top to bottom). $N_{max} = 8$.

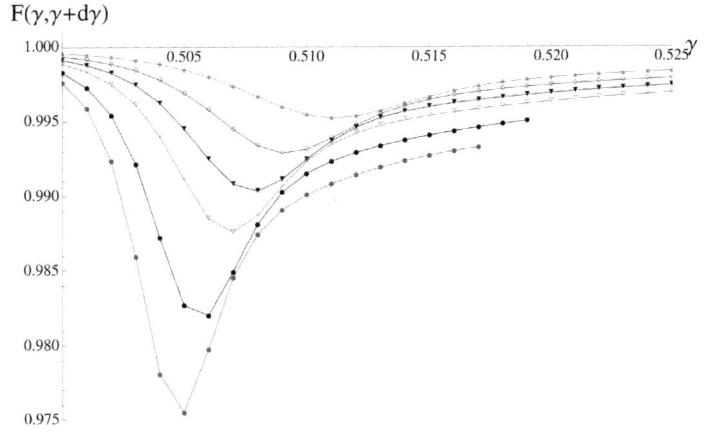

Figure 2. Fidelity as a function of the coupling parameter. With γ from $\gamma_c = 0.5$ to 0.6, $d\gamma = 0.001$, $\omega_0 = \omega = 1$ and $\mathcal{N} = 300$, 400, 500, 600, 800 and 1000 (top to bottom). $N_{max} = 8$.

calculations we use $\omega = \omega_0 = 1$ (resonance) being $\gamma_c = 0.5$ the critical value of the coupling in the thermodynamic limit.

We can locate the coupling's critical value γ_{max}, the value where the quantum phase crossover (the precursor of the QPT) takes place, by identifying the minimum of the fidelity and the maximum of its susceptibility. In figure 5 the value of γ_{max} is shown for each $\mathcal{N}$ in a logarithmic scale. A linear fit gives us

$$\begin{aligned} Log_{10}(\gamma_{max} - \gamma_c) &= -0.285094 - 0.668233\, Log_{10}(\mathcal{N}), \\ (\gamma_{max} - \gamma_c) &= 0.518688\, \mathcal{N}^{-0.668223}. \end{aligned} \tag{11}$$

Where we can obtain the critical exponent $\nu = 0.668223 \simeq 2/3$, which agrees with previous results [16, 17].

In figure 6 the logarithm of the minimum value of the fidelity $log_{10}(F_{min})$ is plotted against the logarithm of the number of atoms. The points call for a quadratic fit, which is:

Journal of Physics: Conference Series **492** (2014) 012012 doi:10.1088/1742-6596/492/1/012012

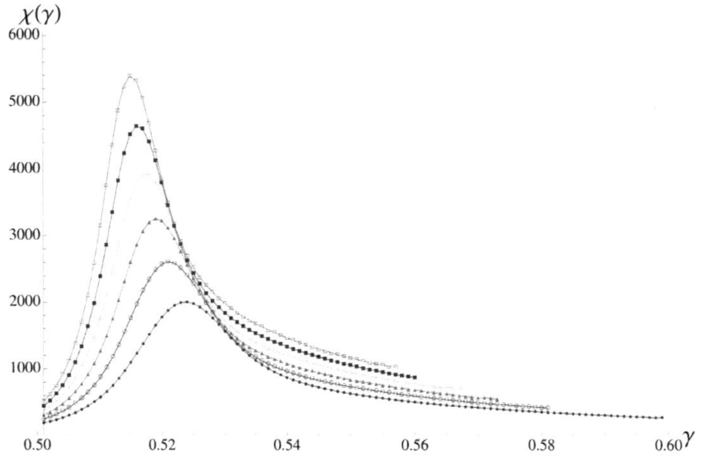

Figure 3. Fidelity susceptibility as a function of the coupling parameter. With γ from $\gamma_c = 0.5$ to 0.6, $d\gamma = 0.001$, $\omega_0 = \omega = 1$ and $\mathcal{N} = 100, 120, 140, 160, 180, 200$ (bottom to top). $N_{max} = 8$.

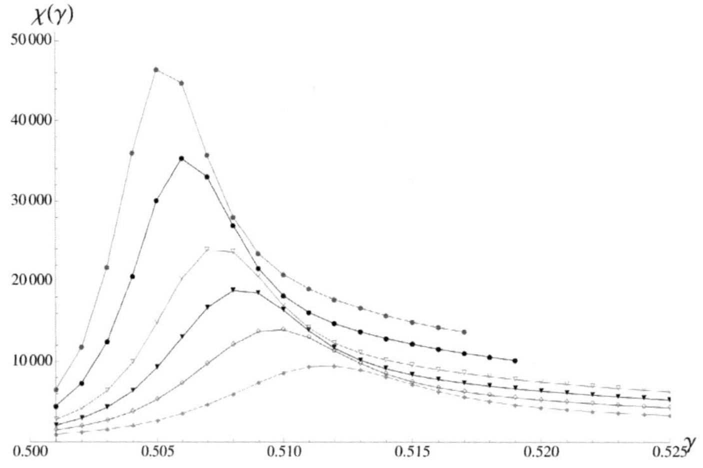

Figure 4. Fidelity susceptibility as a function of the coupling parameter. With γ from $\gamma_c = 0.5$ to 0.6, $d\gamma = 0.001$, $\omega_0 = \omega = 1$ and $\mathcal{N} = 300, 400, 500, 600, 800$ and 1000 (bottom to top). $N_{max} = 8$.

$$Log_{10}(F_{min}) = 0.000351536 - 6.90731 \times 10^{-6}\mathcal{N} - 4.23857 \times 10^{-9}\mathcal{N}^2. \tag{12}$$

We expect that, as we increase the number of atoms, the coefficient of the quadratic term will go to zero. In other words, the quadratic contribution is required by the small $\mathcal{N}$ values, from 100 to 200.

Fig. 7 displays the logarithm of maximum value of the fidelity susceptibility χ^F_{max} as a function of the logarithm of the number of atoms. Fitting linearly the logarithmic curve between the maximum of the susceptibility and $\mathcal{N}$ we obtain:

$$\begin{aligned} Log_{10}\left(\chi^F_{max}\right) &= 0.579291 + 1.36739\, Log_{10}\left(\mathcal{N}\right), \\ \chi^F_{max} &= 3.79569\, \mathcal{N}^{1.36739}. \end{aligned} \tag{13}$$

Journal of Physics: Conference Series **492** (2014) 012012 doi:10.1088/1742-6596/492/1/012012

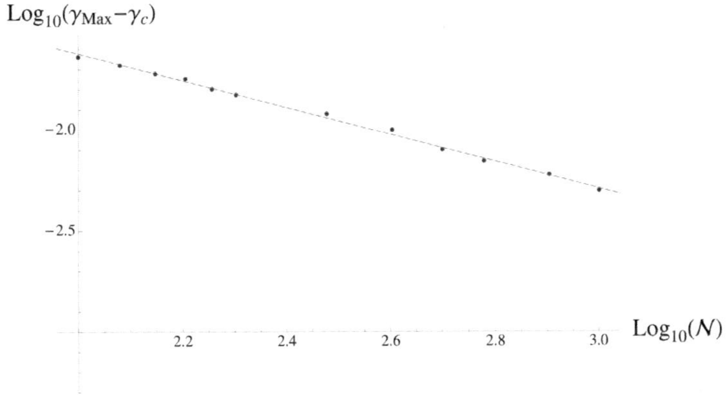

Figure 5. $Log_{10}\left(\gamma_{max}\right)$ as a function of $Log_{10}\left(\mathcal{N}\right)$. With $\mathcal{N}$ from 100 to 1000. $\gamma_c = 0.5$. The linear fit is shown in red.

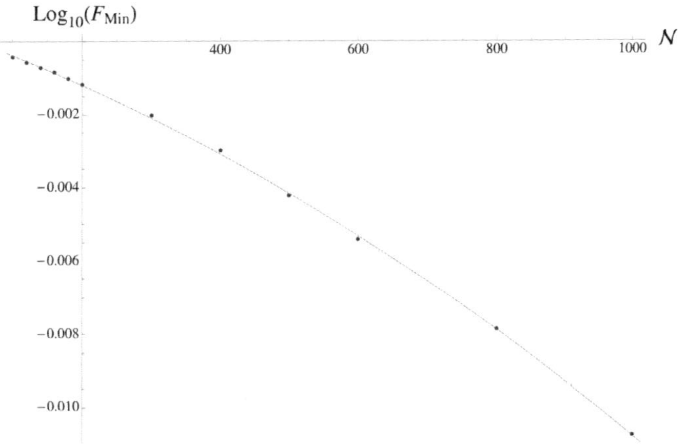

Figure 6. $Log_{10}(F_{min})$ as a function of $\mathcal{N}$. With $\mathcal{N}$ from 100 to 1000. $\gamma_c = 0.5$. A quadratic fit is shown in red.

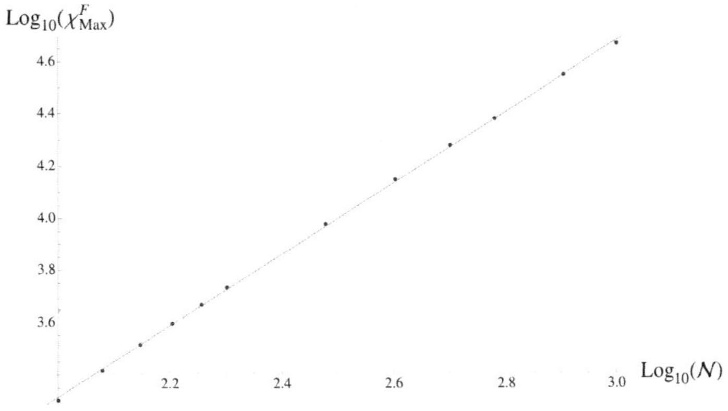

Figure 7. $Log_{10}\left(\chi^{F}_{max}\right)$ as a function of $Log_{10}\left(\mathcal{N}\right)$. With $\mathcal{N}$ from 100 to 1000. $\gamma_c = 0.5$. The linear fit is shown in red.

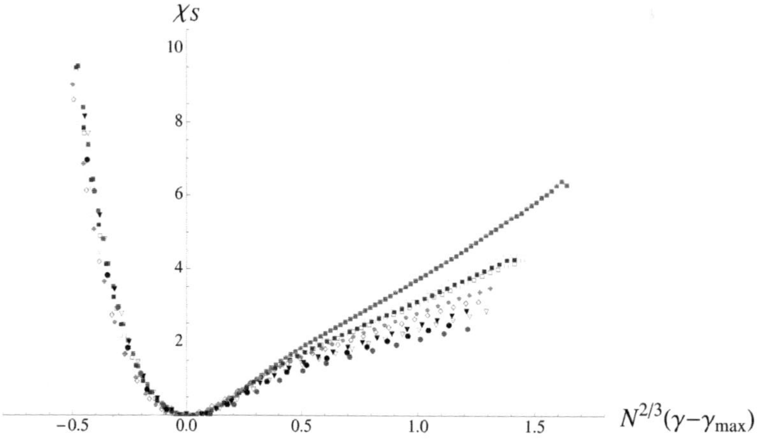

Figure 8. Universal curve for χ_S as a function of $N^{\nu}(\gamma - \gamma_{max})$. With $\mathcal{N}$ from 100 to 1000. $\gamma_c = 0.5$, $\alpha = 2/3$.

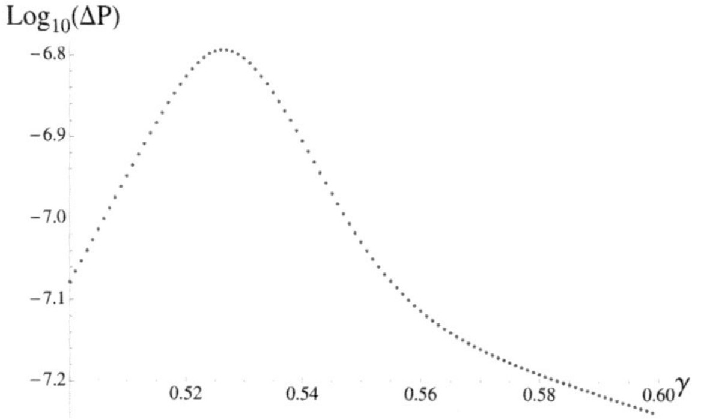

Figure 9. ΔP for the ground state as a function of the coupling parameter. With γ from $\gamma_c = 0.5$ to 0.6, $d\gamma = 0.001$, $\omega_0 = \omega = 1$ and $\mathcal{N} = 100$. $N_{max} = 8$. In this example, the $\gamma_{max} = 0.523$ and the value of the coupling where ΔP has its maximum is $\gamma = 0.526$.

The critical exponent is in this case $1.36739 \simeq 4/3$ which agrees with the one found for the Lipkin-Meshkov-Glick model [23], which belongs to the same universality class [24]. Also, we can calculate the universal curve of the specific susceptibility for every value of $\mathcal{N}$. We show the curve in figure 8. The universal curve guarantees that the critical exponent is correct, because the curves for all $\mathcal{N}$ converge to one curve in the region around the critical value of the coupling strength γ_{max}. The results of figure 8 agree with the ones in [25].

Finally, in figure 9 we show ΔP for the ground state as a function of the coupling for $\mathcal{N} = 100$. As it can be observed, close to the phase transition precursor, which for this number of atoms takes place at $\gamma_{max} = 0.523$, the numerical precision of the ground state wave function becomes smaller. The maximum of this curve occurs at a value of the coupling constant close to, but different from, the γ_{max} calculated through the fidelity and its susceptibility. In all cases the ΔP is small enough to consider that the solution has converged. This behavior repeats for every $\mathcal{N}$. The maximum of the ΔP behaves in a similar way as γ_{max} when the number of atoms grows, taking place closer and closer to γ_c in the thermodynamic limit.

4. Conclusions

We have calculated the fidelity and its susceptibility for the ground state of the finite Dicke model, as functions of the coupling parameter strength, in resonance, for several values of the number of atoms. We located the phase transition for each value of $\mathcal{N}$ using the fidelity formalism, and characterized the phase transition by calculating the critical exponents of the critical values of the coupling and the maximum values of the susceptibility, by fitting logarithmically the curves of both as functions of $\mathcal{N}$. The critical exponents are

$$(\gamma_{max} - \gamma_c) \simeq N^{-0.668223} \simeq N^{-2/3} \text{ and } \chi^F_{max} \simeq N^{1.36739} \sim N^{4/3}. \tag{14}$$

Also, we fitted a quadratic curve of the logarithm of the fidelity as a function of the number of atoms. We validated the values found for the critical exponents plotting the universal curve of the specific susceptibility. Finally, we exhibited that the precision of the ground state wave function (which we use to determine the minimal truncation necessary to have the exact numerical solution) have a maximum near the finite $\mathcal{N}$ phase crossover. Interestingly, those maxima occur at a coupling values slightly different from the ones obtained through the maximum of the fidelity susceptibility .

5. Acknowledgments

This work was partially supported by CONACyT-México and DGAPA-UNAM project IN102811.

6. Appendix

In order to estimate the convergence in the wave function $|\Psi(N_{max})\rangle$, we assume that a similar diagonalization was performed with a truncation $N_{max} - 1$, which provides $|\Psi(N_{max} - 1)\rangle$. To compare both wave functions we extend the latter by assigning $C_{N_{max},m}(N_{max} - 1) = 0$

We define the precision in the calculated wave function as:

$$\begin{aligned}
\Delta P &\equiv 1 - |\langle\Psi(N_{max} - 1)|\Psi(N_{max})\rangle| \\
&= 1 - \left| \sum_{N,N'=0}^{N_{max}} \sum_{m,m'=-j}^{j} C_{N',m'}(N_{max} - 1) C_{N,m}(N_{max}) \langle N'; j, m'|N; j, m\rangle \right| \\
&= 1 - \left| \sum_{N=0}^{N_{max}-1} \sum_{m=-j}^{j} C_{N,m}(N_{max} - 1) C_{N,m}(N_{max}) \right|
\end{aligned}$$

We assume that $N_{max}-1$ is large enough to allow the wave function to be close to convergence. It implies that adding to the Hilbert space the states with N_{max} photon excitations, the components $C_{N,m}, N \leq N_{max} - 1$ will have small changes, conserving their respective phases (but for a global one) with their magnitude remaining constant or slightly decreasing to allow for non-zero $C_{N_{max},m}$ new contributions. This condition can be expressed as

$$|C_{N,m}(N_{max} - 1)| \geq |C_{N,m}(N_{max})|\,, \quad N \leq N_{max} - 1.$$

It follows that

$$\left| \sum_{N=0}^{N_{max}-1} \sum_{m=-j}^{j} C_{N,m}(N_{max} - 1) C_{N,m}(N_{max}) \right| \geq \sum_{N=0}^{N_{max}-1} \sum_{m=-j}^{j} |C_{N,m}(N_{max})|^2,$$

and

$$\Delta P \leq 1 - \sum_{N=0}^{N_{max}-1} \sum_{m=-j}^{j} |C_{N,m}(N_{max})|^2$$
$$= \sum_{m=-j}^{j} |C_{N_{max},m}(N_{max})|^2 .$$

We employ the equality to obtain an upper bound to the precision of the calculated wave functions.

References

[1] Dicke R H 1954 *Phys. Rev.* **93** 99
[2] Scheibner M et al. 2007 *Nature Phys* **3**, 106; Schneble D et al. 2003 *Science* **300**, 475
[3] Hepp K and Lieb E H 1973 *Ann. Phys. (N.Y.)* **76** 360
[4] Wang Y K and Hioe F T 1973 *Phys. Rev.* A **7** 831
[5] Viehmann O, Von Deft J and Marquardt F 2011 *Phys. Rev. Lett.* **107** 113602
[6] Cuiti C and Nataf P 2012 *Phys. Rev. Lett.* **109** 179301
[7] Baumann K, Guerlin C, Brennecke F and Esslinger T 2010 *Nature* **464** 1301
[8] Nagy D, Kónya G, Szirmai G, and Domokos P 2010 *Phys. Rev. Lett.* **104** 130401
[9] Castaños O, Nahmad-Achar E, López Peña R and Hirsch J G 2011 *Phys. Rev.* A **83**051601 (R)
[10] Castaños O, Nahmad-Achar E, López Peña R and Hirsch J G 2011 *Phys. Rev.* A **84** 013819
[11] Emary C and Brandes T 2003 *Phys. Rev.* E **67** 066203; 2003 *Phys. Rev. Lett.* **90**044101
[12] Lambert N, Emary C and Brandes T 2004 *Phys. Rev. Lett.* **92** 073602
[13] Vidal J and Dusuel S 2006 *Europhys. Lett.* **74** 817
[14] Chen Q H, Zhang Y Y, Liu T and Wang K L 2008 *Phys. Rev.* A **78** 051801
[15] Gu S-J 2010 *Int. J. Mod. Phys.* B **24** 4371
[16] Castaños O, López Peña R, Nahmad-Achar E and Hirsch J G 2012 *J. Phys.: Conf. Ser.* **403** 012003
[17] Nahmad-Achar E, Castaños O, López-Peña R and Hirsch J G 2013 *Phys. Scr.* **87** 038114
[18] Leung C-Y, Yu W C, Kwok H-M, Gu S-J and Lin H-Q 2012 *Int. J. Mod. Phys.* B **26** 1250170
[19] Kwok H M, Ning W Q, Gu S J and Lin H Q 2008 *Phys. Rev.* E **78** 032103
[20] Bastarrachea-Magnani M A and Hirsch J G 2011 *Rev. Mex. Fis.* S **57** 0069 `http://rmf.smf.mx/pdf/rmf-s/57/3/57_3_0069.pdf`
[21] Bastarrachea-Magnani M A and Hirsch J G 2012 *AIP Conf. Proc.* **1488** 418
[22] Bastarrachea-Magnani M A and Hirsch J G 2013 *Jour. Phys. A: Math. Gen* submitted
[23] Hirsch J G, Castaños O, López-Peña R, and Nahmad-Achar E 2013 *Phys. Scr.* **87** 038106
[24] Dusuel S and Vidal J 2004 *Phys. Rev. Lett.* **93** 237204; Dusuel S and Vidal J 2005 *Phys. Rev.* **B** 71 224420
[25] Liu T, Zhang Y-Y, Chen Q-H, and Wang K-L 2009 Phys. Rev. A **80** 023810

Journal of Physics: Conference Series **492** (2014) 012013 doi:10.1088/1742-6596/492/1/012013

The Lipkin-Meshkov-Glick model from the perspective of the $SU(1,1)$ Richardson-Gaudin models

Sergio Lerma-H. [1] **and Jorge Dukelsky**[2]

[1] Departamento de Física, Facultad de Física e Inteligencia Artificial, Universidad Veracruzana, Lomas del Estadio s/n, C.P. 91000, Xalapa, Veracruz, México

[2] Instituto de Estructura de la Materia, C.S.I.C., Serrano 123, E-28006 Madrid, Spain

E-mail: slerma@uv.mx

Abstract. Originally introduced in nuclear physics as a numerical laboratory to test different many-body approximation methods, the Lipkin-Meshkov-Glick (LMG) model has received much attention as a simple enough but non-trivial model with many interesting features for areas of physics beyond the nuclear one. In this contribution we look at the LMG model as a particular example of an $SU(1,1)$ Richardson-Gaudin model. The characteristics of the model are analyzed in terms of the behavior of the spectral-parameters or pairons which determine both eigenvalues and eigenfunctions of the model Hamiltonian. The problem of finding these pairons is mathematically equivalent to obtain the equilibrium positions of a set of electric charges moving in a two dimensional space. The electrostatic problems for the different regions of the model parameter space are discussed and linked to the different energy density of states already identified in the LMG spectrum.

1. Introduction

The Lipkin-Meshkov-Glick (LMG) model introduced in the nuclear physics community in a series of three papers in the 60's [1, 2, 3], was originally intended to test different many-body approximation methods and it can be thought as a very schematic model for closed shell nuclei. The model Hamiltonian can be written in terms of the elements of an $SU(2)$ pseudo-spin algebra, and consequently it is an one-quantum-degree of freedom integrable model with the Hamiltonian playing the role of the unique integral of motion. However, simple does not imply trivial and since its introduction different aspects of the model have been studied to gain insights into theoretical concepts such as quantum phase transitions (QPT) [4, 5] and their relations with quantum entanglement properties [6, 7], as well as excited state quantum phase transitions (ESQPT) [8] and quantum decoherence [9]. Moreover, the LMG model has found applications in many other areas of physics like quantum spin systems [10], ion traps [11], Bose-Einstein condensates in double wells [12] or in cavities [13], and in circuit QED [14].

In this contribution we look at the LMG model from the perspective of the Richardson-Gaudin (RG) models for the particular case of a non-compact $SU(1,1)$ algebra. The RG models are Bethe ansatz solvable models which are constructed from a set of L copies of a given Lie algebra. With this set of copies one can construct L independent, quadratic an mutually commuting operators [15]

$$\mathbf{R}_m = \boldsymbol{\xi}_m + \tag{1}$$
$$\sum_{n(n\neq m)}^{L} \left(\sum_{\alpha>0} \frac{|\alpha|^2}{2} \left(X(z_n - z_m)\mathbf{E}_m^{\alpha}\mathbf{E}_n^{-\alpha} + X^*(z_n - z_m)\mathbf{E}_m^{-\alpha}\mathbf{E}_n^{\alpha} \right) + Z(z_n - z_m) \sum_{i=1}^{r} \mathbf{H}_m^i \mathbf{H}_n^i \right),$$

where $\mathbf{H}_m^i$, $\mathbf{E}_m^{\alpha}$, and $\mathbf{E}_m^{-\alpha}$ are the Cartan-Weyl weight, raising and lowering operators of the m-th copy of the given Lie algebra, α are its roots, r its rank, and $\boldsymbol{\xi}_m$ is a linear combination of the weight operators $\boldsymbol{\xi}_m = \sum_{i=1}^{r} k^i \mathbf{H}_m^i$. To ensure the commutativity of the operators, the functions $X(u)$ and $Z(u)$ can take three different forms defining, respectively, the rational, trigonometric and hyperbolic families of the RG models: $X(u) = Z(u) = 1/u$; $X(u) = \csc u, Z(u) = \cot u$; and $X(u) = \mathrm{csch}(u)$, $Z(u) = \coth(u)$. The operators depend on $L + r$ free parameters (z_m and k^i), additionally any function of the $\mathbf{R}_m$ operators defines an integrable model. The number of free parameters allows to link the RG model with a great number of different physical models [16, 17]. The eigenfunctions and eigenvalues of the $\mathbf{R}_m$ can be obtained in terms of a set of spectral parameter or pairons which solve a set of non-linear equations, the so-called Richardson equations. Every solution set of these equations defines a common eigenstate of the operators $\mathbf{R}_m$. It is well known [18] that the set of non-linear Richardson equations is mathematically equivalent to find the equilibrium positions, in a two dimensional plane, of a set of electrical charges (the pairons) in presence of a fixed external electric field or fixed charges (whose positions and values are given by the free parameters z_n and k^i).

In reference [17], it was shown that the LMG model can be obtained as a RG model for the particular case of two copies of an $SU(1,1)$ algebra. In reference [19] this relation was exploited to explore the LMG from the numerical solutions obtained in the RG models. In this contribution some aspects of this relation are discussed, in particular the LMG parameter space is classified in terms of the different families (rational, trigonometric and hyperbolic) of the $SU(1,1)$ RG models. This classification explains naturally the different regimes identified in the energy density of states (EDoS) found in the LMG spectrum [20]. This contribution is organized as follows, in section 2 the explicit relation between the LMG and the $SU(1,1)$ RG models is established. In section 3 the different regimes in the parameter space for the EDoSs of the LMG model are presented and they are linked to the different electrostatic problems associated to the families of the RG models. Conclusions are given in the last section.

2. LMG and $SU(1,1)$ Richardson-Gaudin models

The original physical realization of the LMG model consists of N fermions moving in two N-fold degenerated levels separated by an energy ϵ. In terms of fermionic operators, $\mathbf{a}_{p\sigma}^{\dagger}$ with $p = 1, .., N$ and $\sigma = \pm$, the LMG Hamiltonian reads

$$\mathbf{H}_L = \frac{1}{2}\epsilon \sum_{p\sigma} \sigma \mathbf{a}_{p\sigma}^{\dagger}\mathbf{a}_{p\sigma} + \frac{\lambda}{2} \sum_{pp'\sigma} \mathbf{a}_{p\sigma}^{\dagger}\mathbf{a}_{p'\sigma}^{\dagger}\mathbf{a}_{p'-\sigma}\mathbf{a}_{p-\sigma} + \frac{\gamma}{2} \sum_{pp'\sigma} \mathbf{a}_{p\sigma}^{\dagger}\mathbf{a}_{p'-\sigma}^{\dagger}\mathbf{a}_{p'\sigma}\mathbf{a}_{p-\sigma}.$$

The problem is strongly simplified by noting that the LMG Hamiltonian can be written in terms of $SU(2)$ pseudo-spin operators ($\mathbf{S}_+ = \sum_p \mathbf{a}_{p+}^{\dagger}\mathbf{a}_{p-}$, $\mathbf{S}_- = \sum_p \mathbf{a}_{p-}^{\dagger}\mathbf{a}_{p+}$ and $\mathbf{S}_z = \frac{1}{2}\sum_{p\sigma} \sigma \mathbf{a}_{p\sigma}^{\dagger}\mathbf{a}_{p\sigma}$) in the following form

$$\mathbf{H}_L = \epsilon \mathbf{S}_z + \frac{\lambda}{2}\left(\mathbf{S}_+^2 + \mathbf{S}_-^2\right) + \frac{\gamma}{2}\left(\mathbf{S}_+\mathbf{S}_- + \mathbf{S}_-\mathbf{S}_+\right). \tag{2}$$

From the previous expression it is clear that the invariant subspaces of the LMG Hamiltonian can be classified by the eigenvalues $[j(j+1)]$ of the Casimir operator S^2. For a given pseudo-spin j, two extra invariant subspaces exist which can be labeled by the eigenvalues ($p = \pm 1$)

of the parity operator $\mathbf{P} = e^{i\pi(\mathbf{S}_z+j)}$. The ground state of the model belongs to the subspace with maximal pseudo-spin $j = N/2$ and positive parity. The connection between the LMG Hamiltonian and the $SU(1,1)$ RG models, can be obtained [17] by using the Schwinger boson representation of the $SU(2)$ algebra: $\mathbf{S}_z = \frac{\mathbf{b}_2^\dagger\mathbf{b}_2-\mathbf{b}_1^\dagger\mathbf{b}_1}{2}$, $\mathbf{S}_+ = \mathbf{b}_2^\dagger\mathbf{b}_1$, and $\mathbf{S}_- = \mathbf{b}_1^\dagger\mathbf{b}_2$, where $\mathbf{b}_i$ are boson operators. With these boson operators we construct two copies ($i = 1, 2$) of the rank-1 $SU(1,1)$ algebra

$$\mathbf{K}_i^+ = \frac{1}{2}\mathbf{b}_i^\dagger\mathbf{b}_i^\dagger, \quad \mathbf{K}_i^- = \frac{1}{2}\mathbf{b}_i\mathbf{b}_i, \quad \mathbf{K}_i^z = \frac{1}{2}\left(\mathbf{b}_i^\dagger\mathbf{b}_i + \frac{1}{2}\right), \tag{3}$$

where the ladder $(\mathbf{K}^+, \mathbf{K}^-)$ and weight $(\mathbf{K}^z)$ operators satisfy the commutation relations $[\mathbf{K}^z, \mathbf{K}^\pm] = \pm\mathbf{K}^\pm$ and $[\mathbf{K}^+, \mathbf{K}^-] = -2\mathbf{K}^z$. The only root of the algebra is $|\alpha|^2 = 2$ and the Cartan Weyl basis is given by $\mathbf{H}_1 = \sqrt{2}\mathbf{K}^z$, $\mathbf{E}^\alpha = \mathbf{K}^+$ and $\mathbf{E}^{-\alpha} = -\mathbf{K}^-$. With these two copies and choosing $k^1 = 1/(\sqrt{2}g)$ in Eq.(1), we construct two integrals of motion

$$g\mathbf{R}_i = \mathbf{K}_i^z + g\sum_{j\neq i}\left[-X_{ij}\left(\mathbf{K}_i^+\mathbf{K}_j^- + \mathbf{K}_i^-\mathbf{K}_j^+\right) + 2Z_{ij}\mathbf{K}_i^z\mathbf{K}_j^z\right], \tag{4}$$

where $i = 1, 2$, and we have reparametrized [17] the functions $X(z_i - z_j) = X_{ij}$ and $Z(z_i - z_j) = Z_{ij}$ in the way $X_{21} = -X_{12} = \frac{1+st^2}{2t}$ and $Z_{21} = -Z_{12} = \frac{1-st^2}{2t}$, with $s = -1, 0, 1$ corresponding, respectively, to the hyperbolic, rational and trigonometric versions of the $SU(1,1)$ RG models. The parameters t and g are completely free parameters. By taking the difference between the previous integrals of motion, we obtain the LMG Hamiltonian

$$\mathbf{H}_L = \epsilon(g\mathbf{R}_2 - g\mathbf{R}_1) - \frac{\gamma}{4}, \qquad \text{with} \qquad gX_{21} = -\frac{\lambda}{\epsilon}, \quad \text{and} \quad gZ_{21} = \frac{\gamma}{\epsilon}. \tag{5}$$

The relation between the LMG Hamiltonian parameters (λ, γ, ϵ) and those of the $SU(1,1)$ RG models are $\frac{\lambda}{\epsilon} = -g\frac{1+st^2}{2t}$ and $\frac{\gamma}{\epsilon} = g\frac{1-st^2}{2t}$. In order to obtain a simpler parameter space, we introduce the rescaled parameters [4] $(\gamma_x, \gamma_y) \equiv \frac{2j-1}{\epsilon}(\gamma + \lambda, \gamma - \lambda) = (2j-1)g\left(-st, \frac{1}{t}\right)$. This relation allows to classify the quadrants of the LMG parameter space of Fig.1 in terms of the hyperbolic ($s = -1$) and trigonometric ($s = 1$) RG models. The first ($s = -1$, $g > 0$) and third ($s = -1$, $g < 0$) quadrants correspond to the hyperbolic RG model, whereas the second ($s = 1$, $g > 0$) and fourth ($s = 1$, $g < 0$) are associated with the trigonometric model. The eigenvalues of the LMG model in terms of the pairons (e_α) of the RG model, for pseudo-spin j and positive parity, are given by $E_L = gt\epsilon\sum_\alpha^j \frac{1+se_\alpha^2}{t^2-e_\alpha^2}$. The unnormalized eigenvectors common to the two integrals of motion ($\mathbf{R}_i$) and, consequently, to the LMG Hamiltonian are

$$\prod_{\alpha=1}^{j}\left(\frac{\mathbf{b}_1^\dagger\mathbf{b}_1^\dagger}{e_\alpha + t} + \frac{\mathbf{b}_2^\dagger\mathbf{b}_2^\dagger}{e_\alpha - t}\right)|0\rangle, \tag{6}$$

with $|0\rangle$ the boson vacuum. The equations which determine the pairons and their mapping to a two-dimensional electrostatic problem for the different version of the RG models are discussed in the next section. Likewise the electrostatic systems are linked to the different regimes for the EDoS already found in [20] for the LMG energy spectrum.

3. Energy Density for the LMG states and the RG electrostatic mapping

In figure 1 the parameter space of the LMG model is classified according to the different EDoSs that can be found in the LMG spectrum. Typical profiles of these densities are shown for the

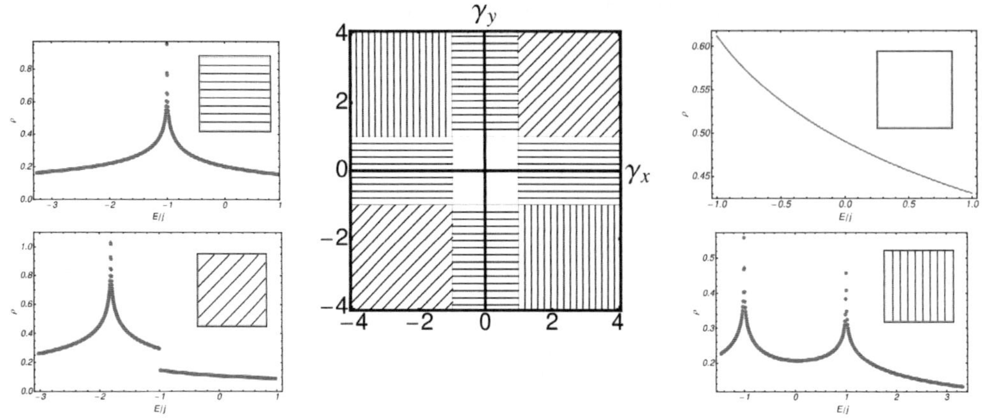

Figure 1. Central plot: parameter space of the LMG model classified according to the different structures that can be found in the Energy Density of States (horizontal, vertical, diagonal and white regions). Typical profiles of these densities (ρ) are shown in the plots surrounding the central one, the corresponding regions are indicated by the squares inside the plots. The density profiles were calculated numerically for the positive sector of finite ($j = 400$) LMG Hamiltonians. The first and third quadrants in the central parameter space correspond to the hyperbolic RG model, whereas the second and fourth are associated with the trigonometric RG family.

different regions. These densities were calculated numerically for the parity positive sector of finite ($j = 400$) LMG systems. The white region around the non-interacting ($\gamma_x = \gamma_y = 0$) case shows monotone EDoSs, with energies in the interval $E \in [-j, j]$. As we move away from the origin a transition to the regions signaled by horizontal lines takes place. These regions are characterized by a sole peak in the EDoS occurring at $E/j = -1$ in the left and bottom horizontal lined regions, and at $E/j = 1$ in the right and top horizontal lined ones. This peak signals an excited state quantum phase transition [21, 8] which transform in an ordinary (Ground-state) quantum phase transition in the left and bottom border of the white region. Away from the axis of the parameter space, two types of regions can be identified. The first type signaled by vertical lines in the second and fourth quadrants of the parameters space, is characterized by two peaks in the EDoS occurring at $E = -j$ and $E = +j$. The double peak structure of the EDoS is related to excited state quantum phase transitions which evolve, respectively, from the ground and from the most excited state. Finally, the second type of regions (signaled by diagonal lines in Figure 1) in the first and third quadrants of the parameter space, presents EDoSs with a peak and a discontinuity. This discontinuity occurs at $E = -j$ in the left bottom diagonal lined region, and at $E = j$ in the other one. The states with energies between the peak and the discontinuity present avoiding crossings, whereas the states beyond the discontinuity show a very small energy density. The qualitative differences between the EDoSs of the first-third and second-fourth quadrants found a natural explanation when we look at the equivalent electrostatic problem which determine the pairons of the RG solution. These pairons determine, in turn, the wave function and energies of the LMG Hamiltonian as it was discussed in the previous section. The equations determining the pairons can be written [19] in the following form, where it is clear their analogy with an electrostatic problem in the complex plane

$$\frac{Q_C}{e_\alpha - P_C} + \frac{Q_D}{e_\alpha - P_D} + \frac{1}{4}\left(\frac{1}{e_\alpha + t} + \frac{1}{e_\alpha - t}\right) + \sum_{\beta \neq \alpha}^{M} \frac{1}{e_\alpha - e_\beta} = 0. \tag{7}$$

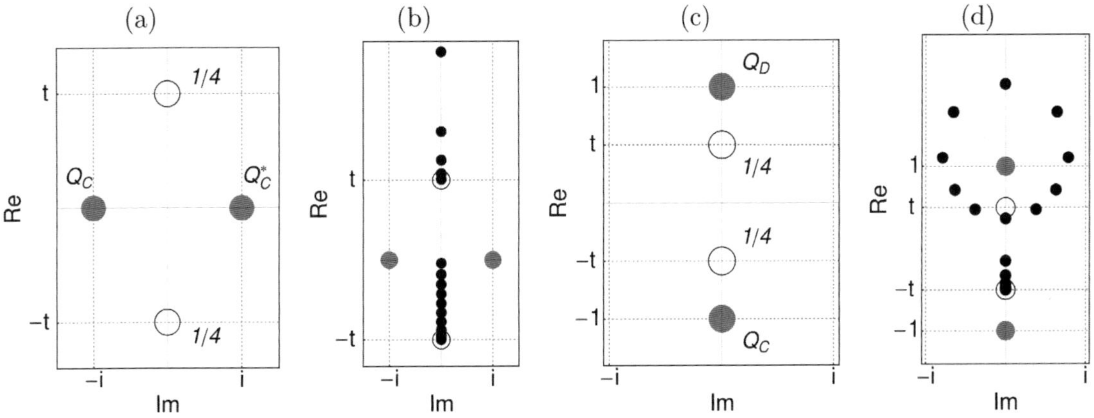

Figure 2. Fixed charges of the equivalent electrostatic problems in the complex plane for the trigonometric (panel a) and hyperbolic (panel c) RG models. The effective charges Q_C and Q_D are complex conjugated for the trigonometric case, whereas in the hyperbolic case they are real. The other two effective charges have a value 1/4 and they are located in the real axis at $\pm t$ for both hyperbolic and trigonometric cases. Two typical equilibrium configurations for two LMG excited states are shown in panel b and c for the trigonometric and hyperbolic cases respectively. A value $j = 15$ was used.

The previous equations describe the electrostatic interaction of a set of j pairons (whose positions are e_α) with positive unit charge in a two dimensional space. The first two terms in (7) describe the electrostatic interaction of the pairons with two effective charges Q_C and Q_D. For the trigonometric case ($s = 1$) the effective charges are complex $Q_C = -\frac{2j-1}{4} - \frac{i}{4g}$, $Q_D = Q_C^*$, and they are located in $P_C = i$, $P_D = -i$. Whereas in the hyperbolic case ($s = -1$) both, the charges and their positions, are real, $Q_C = -\frac{2j-1}{4} + \frac{1}{4g}$ and $Q_D = -\frac{2j-1}{4} - \frac{1}{4g}$, located in $P_C = -1$ and $P_D = 1$ respectively. The third term in (7) represents the interaction of the pairons with two charges $\frac{1}{4}$ at positions $\mp t$. Finally, the fourth term corresponds to the mutual repulsion between pairons.

The panels (a) and (c) of Fig. 2 show the position of the effective fixed charges for the trigonometric and hyperbolic RG families. Each independent solution of the Richardson equations determines the equilibrium position of the pairons in the complex plane. For the trigonometric case, the dynamics of the pairons as a function of the model parameters takes place entirely in the real axis (panel (b) of Fig.2). This result can be understood from the electrostatic mapping as a consequence of the different sign of the imaginary part of the effective charges Q_C and $Q_D = Q_C^*$. These imaginary parts produce effective electric fields whose components parallel to the imaginary axis cancel mutually only in the real axis, constraining the equilibrium positions of the pairons to this axis. Contrarily to this, for the hyperbolic case the effective charges Q_C and Q_D are real and located at the real axis at positions ± 1, moreover the mutual sign of these charges can be different depending on the values of the LMG parameters. When the mutual sign of these charges is negative, the pairons can expand in the complex plane forming arcs around the position of the effective charges Q_C or Q_D. Panel (d) of Fig.2 shows a such configuration with an arc around the Q_D charge. For particular values of the coupling constant [$g = \pm 1/(2j + 1 - 2n)$, with n a positive integer] n pairons can collapse into the position of the effective charge Q_C or Q_D. Similar collapses were found in the $SU(2)$-RG-hyperbolic model for the $p_x + ip_y$ pairing interaction [22]. In that model the collapses of the pairons were associated with a third-order QPT. In the present model the collapses are associated with other singular behavior: the crossing of different parity states at exactly the same values of the coupling constant where collapses take place. The qualitative differences between the electrostatic systems for the trigonometric (second and third quadrants) and hyperbolic (first and third quadrants) RG models, explain qualitatively the different EDoSs found in the LMG parameter space of Fig. 1.

4. Conclusions

The Lipkin-Meshkov-Glick model was studied from the perspective of a particular realization of the Richardson-Gaudin models, associated with two bosonic copies of an $SU(1,1)$ algebra. In the parameter space of the LMG model different profiles of the energy density of states can be identified. The different profiles were related with the kind of electrostatic problem that determines the pairons or spectral parameters of the RG models. The EDoSs with a peak and a discontinuity are associated with the hyperbolic RG models, for which the pairon dynamics as a function of the model parameters takes place in the whole complex plane. Contrarily, for the trigonometric RG family, where the electrostatic pairon positions are restricted to the real axis, the EDoSs present a structure with two peaks at energies $E = -j$ and $E = j$. As discussed in Ref.[23], the pairon dynamics can help to identify significant physical phenomena. The LMG model is not an exception, for the hyperbolic case the pairon dynamics allows the collapse of pairons into the position of an effective charge. This singular behavior is related with the crossing between different parity states and the avoiding crossing between states of the same parity. Finally, the insights gained in the study of the LMG model, where the set of pairons of every state in the spectrum is easily accessible, can help to identify general mechanisms related with ubiquitous phenomena such as excited state quantum phase transitions.

Acknowledgments

J. D. acknowledges support from project FIS2012-34479 of the Ministerio de Economía y Competitividad (Spain).

References

[1] Lipkin H J, Meshkov N and Glick A J 1965 *Nucl. Phys.* **62** 188
[2] Lipkin H J, Meshkov N and Glick A J 1965 *Nucl. Phys.* **62** 199
[3] Lipkin H J, Meshkov N and Glick A J 1965 *Nucl. Phys.* **62** 211
[4] Castaños O, López-Peña R, Hirsch J and López-Moreno E 2006 *Phys. Rev.* B **74** 104118
[5] Vidal J, Arias J M, Dukelsky J and García-Ramos J E 2006 *Phys. Rev.* C **73** 054305
[6] Vidal J, Palacios G and Mosseri R 2004 *Phys. Rev.* A **69** 022107
[7] Vidal J, Dukelsky J and Mosseri R 2004 *Phys. Rev.* A **69** 054101
[8] Heiss W D, Scholtz F G and Geyer H B 2005 *J. Phys.* A **38** 1843
[9] Relaño A, Arias J M, Dukelsky J, García-Ramos J E and Pérez-Fernández P 2008 *Phys. Rev.* A **78** 060102
[10] Botet R, Julien R and Pfeuty P 1982 *Phys. Rev. Lett.* **49** 478
[11] Unanyan R G and Fleischhauer M 2003 *Phys. Rev. Lett.* **90** 133601
[12] Links J, Zhou H Q, McKenzie R H and Gould M D 2003 *J. Phys.* A **36** R63
[13] Chen G, Liang J Q and Jia S 2009 *Opt. Express* **17** 19682
[14] Larson J 2010 *Europhys. Lett.* **90** 54001
[15] Lerma H. S and Errea B 2007 *J. Phys.* A **40** 4125
[16] Dukelsky J, Pittel S and Sierra G 2004 *Rev. Mod. Phys.* **76** 643
[17] Ortiz G, Somma R, Dukelsky J and Rombouts S 2005 *Nucl. Phys.* B **707** 421
[18] Dukelsky J, Esebbag C and Pittel S 2002 *Phys. Rev. Lett.* **88** 062501
[19] Lerma H. S and Dukelsky J 2013 *Nucl. Phys.* B **870** 421
[20] Ribeiro P, Vidal J, Mosseri R 2007 *Phys. Rev. Lett.* **99** 050402
[21] Perez-Fernández P, Relaño A, Arias J M, Cejnar P, Dukelsky J and García-Ramos J E 2011 *Phys. Rev.* E **83** 046208
[22] Ibañez M, Links J, Sierra G and Zhao S-Y, 2009 *Phys. Rev.* B **79** 180501
[23] Rubeni D, Foerster A, Mattei E and Roditi I 2012 *Nucl. Phys.* B **856** 698

Multi-dimensional potential energy surfaces and non-axial octupole correlations in actinide and transfermium nuclei from relativistic mean field models

Bing-Nan Lu[1], Jie Zhao[1], En-Guang Zhao[1,2], Shan-Gui Zhou[1,2]

[1]State Key Laboratory of Theoretical Physics, Institute of Theoretical Physics, Chinese Academy of Sciences, Beijing 100190, China

[2]Center of Theoretical Nuclear Physics, National Laboratory of Heavy Ion Accelerator, Lanzhou 730000, China

E-mail: sgzhou@itp.ac.cn (Shan-Gui Zhou)

Abstract. We have developed multi-dimensional constrained covariant density functional theories (MDC-CDFT) for finite nuclei in which the shape degrees of freedom $\beta_{\lambda\mu}$ with even μ, e.g., β_{20}, β_{22}, β_{30}, β_{32}, β_{40}, etc., can be described simultaneously. The functional can be one of the following four forms: the meson exchange or point-coupling nucleon interactions combined with the non-linear or density-dependent couplings. For the pp channel, either the BCS approach or the Bogoliubov transformation is implemented. The MDC-CDFTs with the BCS approach for the pairing (in the following labelled as MDC-RMF models with RMF standing for "relativistic mean field") have been applied to investigate multi-dimensional potential energy surfaces and the non-axial octupole Y_{32}-correlations in $N = 150$ isotones. In this contribution we present briefly the formalism of MDC-RMF models and some results from these models. The potential energy surfaces with and without triaxial deformations are compared and it is found that the triaxiality plays an important role upon the second fission barriers of actinide nuclei. In the study of Y_{32}-correlations in $N = 150$ isotones, it is found that, for ^{248}Cf and ^{250}Fm, $\beta_{32} > 0.03$ and the energy is lowered by the β_{32} distortion by more than 300 keV; while for ^{246}Cm and ^{252}No, the pocket with respect to β_{32} is quite shallow.

1. Introduction

The shape of a nucleus can be described by the parametrization of the nuclear surface or the nucleon density distribution [1, 2]. One of the mostly used parametrizations is the multipole expansion with coefficients $\beta_{\lambda\mu}$'s; see Fig. 1 [3] for a schematic show of some typical nuclear shapes. Many shape degrees of freedom are important not only for the ground states or small amplitude collective motions, but also for large amplitude collective motions such as fission. The readers are referred to Ref. [4] for an overview. Here we simply mention that both the non-axial and the reflection symmetries should be broken in the study of nuclear ground state properties and potential energy surfaces.

We have developed multi-dimensional constrained covariant density functional theories (MDC-CDFT) by breaking the reflection and the axial symmetries simultaneously [3–7]. In these theories, the nuclear shape is assumed to be invariant under the reversion of x and y axes,

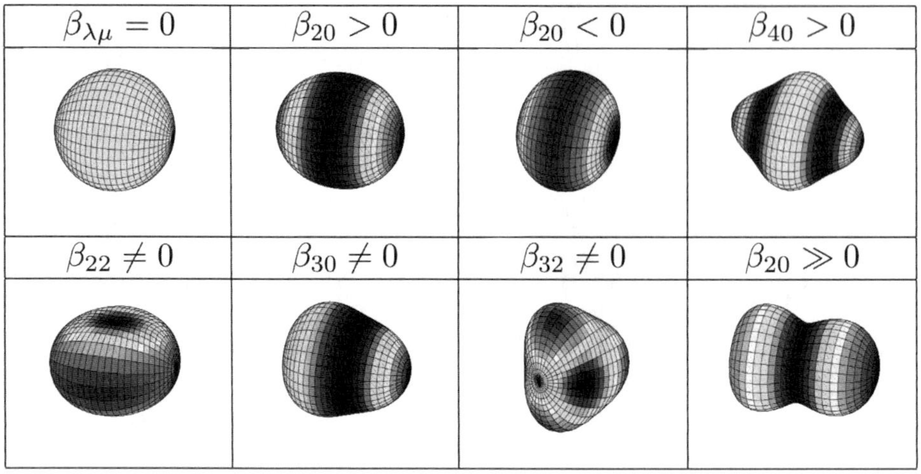

Figure 1. (Color online) A schematic show of some typical nuclear shapes. From left to right, the 1st row: (a) Sphere, (b) Prolate spheroid, (c) Oblate spheroid, (d) Hexadecapole shape, and the second row: (e) Triaxial ellipsoid, (f) Reflection symmetric octupole shape, (g) Tetrahedron, (h) Reflection asymmetric octupole shape with very large quadrupole deformation and large hexadecapole deformation. Taken from Ref. [3].

i.e., the intrinsic symmetry group is V_4 and all shape degrees of freedom $\beta_{\lambda\mu}$ with even μ, e.g., β_{20}, β_{22}, β_{30}, β_{32}, β_{40}, $\cdots$, are included self-consistently. The covariant density functional can be one of the following four forms: the meson exchange or point-coupling nucleon interactions combined with the non-linear or density-dependent couplings. For the pp channel, either the BCS approach or the Bogoliubov transformation is implemented. The MDC-CDFT with the BCS approach for the pairing is named as MDC-RMF and that with the Bogoliubov transformation as MDC-RHB [4]. In this contribution, we will present the formalism for MDC-RMF models and some results of actinide nuclei.

In Section 2, we give the formalism of MDC-RMF models briefly. The results for actinide nuclei are shown and discussed in Section 3. A summary is given in Section 4.

2. Formalism of MDC-RMF models

In this section, we briefly give the formalism of MDC-RMF models. More details can be found in Ref. [4]. The starting point of a RMF model with the non-linear point coupling interactions is the following Lagrangian [8–16],

$$\mathcal{L} = \bar{\psi}\left(i\gamma_\mu\partial^\mu - M_B\right)\psi - \mathcal{L}_{\mathrm{lin}} - \mathcal{L}_{\mathrm{nl}} - \mathcal{L}_{\mathrm{der}} - \mathcal{L}_{\mathrm{cou}}, \tag{1}$$

where

$$\begin{aligned}
\mathcal{L}_{\mathrm{lin}} &= \frac{1}{2}\alpha_S\rho_S^2 + \frac{1}{2}\alpha_V j_V^2 + \frac{1}{2}\alpha_{TS}\vec{\rho}_{TS}^2 + \frac{1}{2}\alpha_{TV}\vec{j}_{TV}^2, \\
\mathcal{L}_{\mathrm{nl}} &= \frac{1}{3}\beta_S\rho_S^3 + \frac{1}{4}\gamma_S\rho_S^4 + \frac{1}{4}\gamma_V\left[j_V^2\right]^2, \\
\mathcal{L}_{\mathrm{der}} &= \frac{1}{2}\delta_S\left[\partial_\nu\rho_S\right]^2 + \frac{1}{2}\delta_V\left[\partial_\nu j_V^\mu\right]^2 + \frac{1}{2}\delta_{TS}\left[\partial_\nu\vec{\rho}_{TS}\right]^2
\end{aligned}$$

$$
\begin{aligned}
& +\frac{1}{2}\delta_{TV}\left[\partial_\nu \vec{j}_{TV}\right]^2, \\
\mathcal{L}_{\rm cou} &= \frac{1}{4}F^{\mu\nu}F_{\mu\nu} + e\frac{1-\tau_3}{2}A_\mu j_V^\mu, \qquad (2)
\end{aligned}
$$

are the linear, non-linear, and derivative couplings and the Coulomb part, respectively. M_B is the nucleon mass, α_S, α_V, α_{TS}, α_{TV}, β_S, γ_S, γ_V, δ_S, δ_V, δ_{TS}, and δ_{TV} are coupling constants for different channels and e is the electric charge. ρ_S, $\vec{\rho}_{TS}$, j_V, and $\vec{j}_{TV}$ are the iso-scalar density, iso-vector density, iso-scalar current, and iso-vector current, respectively.

Starting from the above Lagrangian and under several approximations, one can derive the Dirac equation for the nucleons,

$$\hat{h}\psi_i = \{\alpha\cdot\mathbf{p} + \beta[M_B + S(\mathbf{r})] + V(\mathbf{r})\}\,\psi_i = \epsilon_i\psi_i, \qquad (3)$$

where the potentials $V(\mathbf{r})$ and $S(\mathbf{r})$ are calculated from the densities.

An axially deformed harmonic oscillator (ADHO) basis is adopted for solving the Dirac equation [3–7, 17–19]. Note that a RMF model with reflection asymmetry has been developed in a two-center HO basis [20]. The ADHO basis are defined as the eigen solutions of the Schrödinger equation with an ADHO potential,

$$\left[-\frac{\hbar^2}{2M_B}\nabla^2 + \frac{1}{2}M_B(\omega_\rho^2\rho^2 + \omega_z^2 z^2)\right]\Phi_\alpha(\mathbf{r}\sigma) = E_\alpha\Phi_\alpha(\mathbf{r}\sigma), \qquad (4)$$

where ω_z and ω_ρ are the oscillator frequencies along and perpendicular to the z axis, respectively. These basis are also eigen functions of the z component of the angular momentum j_z with eigen values $K = m_l + m_s$. For any basis state $\Phi_\alpha(\mathbf{r}\sigma)$, the time reversal state is defined as $\Phi_{\bar{\alpha}}(\mathbf{r}\sigma) = \mathcal{T}\mathbf{\Phi}_\alpha(\mathbf{r}\sigma)$, where $\mathcal{T} = i\sigma_y K$ is the time reversal operator and K is the complex conjugation. Apparently we have $K_{\bar{\alpha}} = -K_\alpha$ and $\pi_{\bar{\alpha}} = \pi_\alpha$ with $\pi = \pm 1$ being the parity. These basis form a complete set for expanding any two-component spinors. For a Dirac spinor with four components,

$$\psi_i(\mathbf{r}\sigma) = \begin{pmatrix} \sum_\alpha f_i^\alpha \Phi_\alpha(\mathbf{r}\sigma) \\ \sum_\alpha g_i^\alpha \Phi_\alpha(\mathbf{r}\sigma) \end{pmatrix}, \qquad (5)$$

where the sum runs over all the possible combination of the quantum numbers $\alpha = \{n_z, n_r, m_l, m_s\}$ and f_i^α and g_i^α are the expansion coefficients. In practical calculations, one should truncate the basis in an effective way.

In our model the nucleus is assumed to be symmetric under the V_4 group, that is, for all the potentials and densities we can do the Fourier series expansion,

$$f(\rho,\varphi,z) = f_0(\rho,z)\frac{1}{\sqrt{2\pi}} + \sum_{n=1}^{\infty} f_n(\rho,z)\frac{1}{\sqrt{\pi}}\cos(2n\varphi), \qquad (6)$$

where $f_0(\rho, z)$ and $f_n(\rho, z)$ are real functions of ρ and z. The formalism for calculating the Fourier components of the potentials and densities can be found in Ref. [4].

Either the BCS approach or the Bogoliubov transformation has been implemented in our model to take into account the pairing effects. For the pairing force, we can use a delta force or a separable finite-range pairing force [21–23].

To obtain a potential energy surface, i.e., the energy of a nucleus as a function of deformation parameters, we make multi-dimensional constraint calculations which are equivalent to adding external potentials during the iteration. A modified linear constraint method was included in our MDC-RMF calculations [3–7].

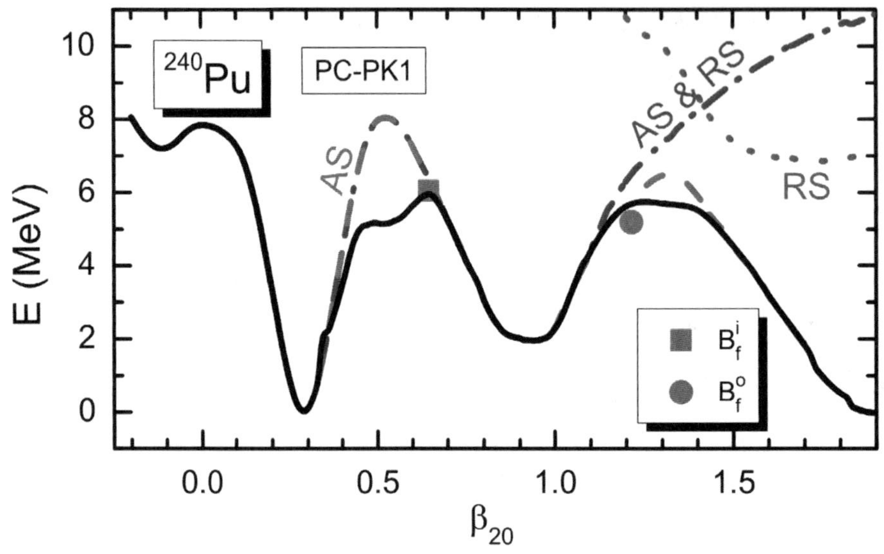

Figure 2. (Color online) Potential energy curve of ^{240}Pu with various self-consistent symmetries imposed. The solid black curve represents the calculated fission path with V_4 symmetry imposed, the red dashed curve that with the axial symmetry (AS) imposed, the green dotted curve that with the reflection symmetry (RS) imposed, the violet dot-dashed line that with both symmetries (AS & RS) imposed. The empirical inner (outer) barrier height is taken from Ref. [24] and denoted by the grey square (circle). The energy is normalized with respect to the binding energy of the ground state. The parameter set used is PC-PK1. Taken from Ref. [5].

The total energy of a nucleus is obtained by substituting the densities into the expectation value of the Hamiltonian. The center of mass correction $E_{\text{c.m.}}$ can be calculated either phenomenologically or microscopically, depending on the effective interactions. The intrinsic multipole moments are calculated from the vector densities by

$$Q_{\lambda\mu} = \int d^3\mathbf{r}\rho_V(\mathbf{r})r^\lambda Y_{\lambda\mu}(\Omega), \tag{7}$$

where $Y_{\lambda\mu}(\Omega)$ is the spherical harmonics.

3. Results and discussions

In Ref. [5], one- (1-d), two- (2-d), and three-dimensional (3-d) constraint calculations were made for the actinide nucleus ^{240}Pu using the MDC-RMF models with the parameter set PC-PK1 [25, 26]. In Fig. 2 we show the 1-d potential energy curve from an oblate shape with β_{20} about -0.2 to the fission configuration with β_{20} beyond 2.0 which are obtained from calculations with different self-consistent symmetries imposed: The axial (AS) or triaxial (TS) symmetries combined with reflection symmetric (RS) or asymmetric cases. The importance of the triaxial deformation on the inner barrier [27, 28] and that of the octupole deformation on the outer barrier are clearly seen: The triaxial deformation reduces the inner barrier height by more than 2 MeV and results in a better agreement with the empirical value [24]; the RA shape is favored beyond the fission isomer and lowers very much the outer fission barrier. Besides these features, it was found for the first time that the outer barrier is also considerably lowered by about 1 MeV when the triaxial deformation is allowed. In addition, a better reproduction of the empirical

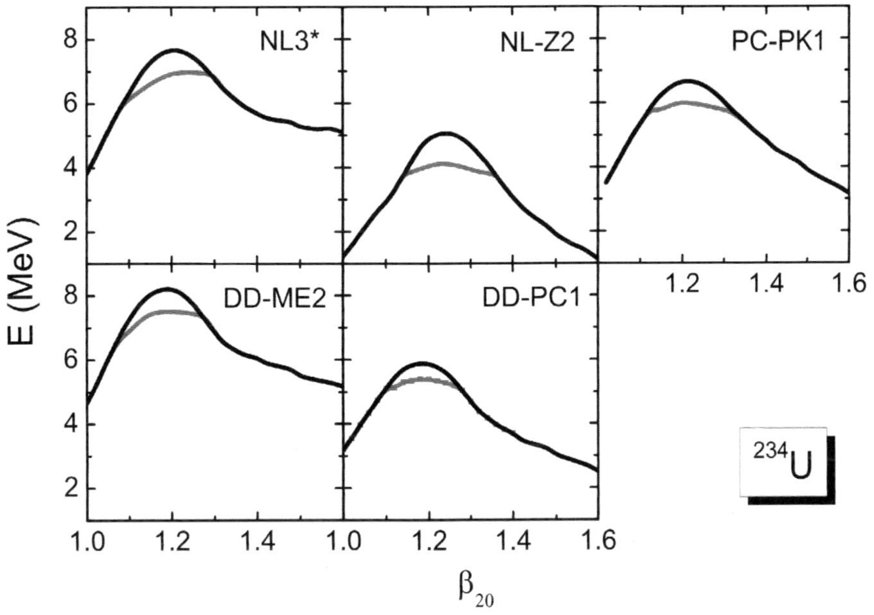

Figure 3. (Color online) Potential energy curves of ^{234}U in the second barrier region calculated from MDC-RMF models with the NL3* [29, 30], NL-Z2 [31], PC-PK1 [25, 26], DD-ME2 [32], and DD-PC1 [33] parameter set, respectively. The reflection asymmetric shapes are allowed. The axial and non-axial symmetric results are denoted by black and red curves, respectively. Taken from Ref. [3].

barrier height can be seen for the outer barrier. It has been stressed that this feature can only be found when the axial and reflection symmetries are simultaneously broken [5].

In Ref. [5], it was also examined the parameter dependence of the influence of triaxiality on the second fission barrier and the lowering effect of the triaxiality on the second fission barrier was also observed when parameter sets other than PC-PK1 are used. In Fig. 3 we show the comparison of the potential energy curves of ^{234}U in the second barrier region calculated from the MDC-RMF models with different parameter sets and Lagrangian forms, including meson-exchange ones NL3* [29, 30], NL-Z2 [31], DD-ME2 [32], and point-coupling ones PC-PK1 [25, 26] and DD-PC1 [33]. The results with and without triaxiality are both presented. From the figure we can see that, although the absolute values of the barriers differ a lot among these results, for all five parameter sets the second barriers are lowered by the triaxiality. The largest effect is around 1 MeV for NL-Z2 and the smallest one is around 300 keV for DD-PC1. Thus we conclude that the lowering effects of the triaxiality on the second barriers are parameter independent in the RMF models.

The self-consistent three-dimensional constraint calculations are very time-consuming. From the benchmark calculations for ^{240}Pu, we learned many experiences about the important roles played by various shape degrees of freedom in different regions of the deformation space. For example, around the first fission barrier an actinide nucleus assumes triaxial and reflection symmetric shapes but around the second fission barrier both triaxial and octupole deformations are important [4]. These experiences are used in a systematic study of even-even actinide nuclei and the results were presented in Ref. [4]. Here in Fig. 4 we show the 1-d potential energy curves for even-even actinide nuclei from the fission isomer with $\beta_{20} = 1.0$ to the fission configuration with $\beta_{20} = 1.7$ calculated from the MDC-RMF model with the parameter set PC-

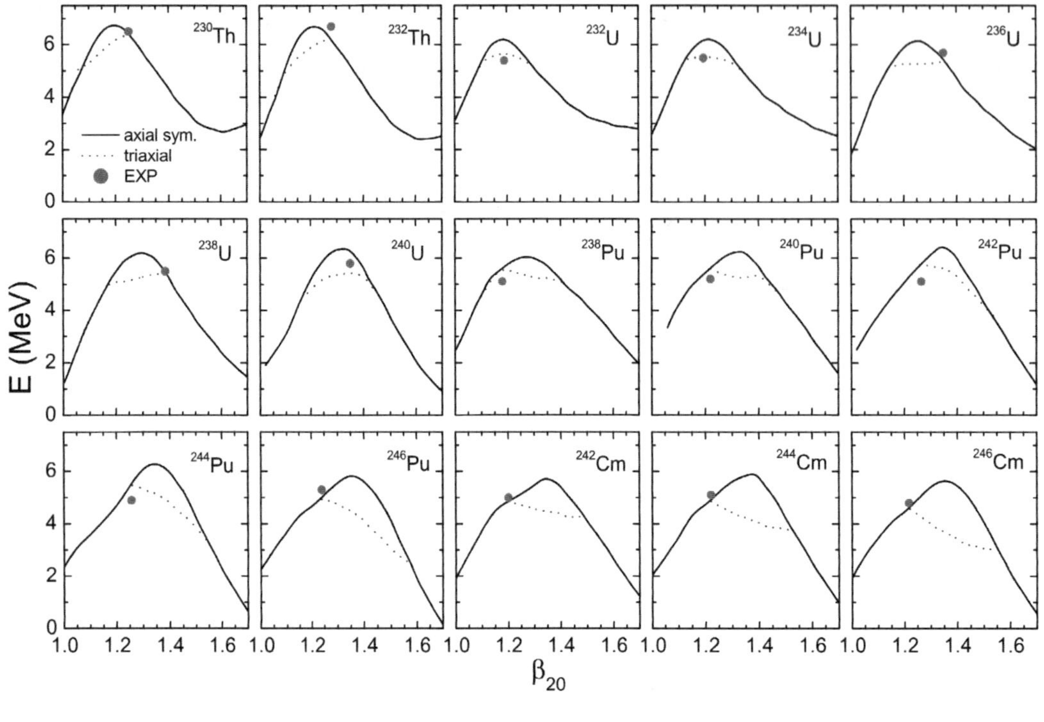

Figure 4. (Color online) Potential energy curves of even-even actinides nuclei in the second barrier regions calculated from the MDC-RMF model with the PC-PK1 parameter set [25, 26]. The reflection asymmetric shapes are allowed. The axial and non-axial symmetric results are denoted by solid and dotted curves, respectively. The binding energies are normalized with respect to the ground states. The empirical values are taken from Ref. [24] and denoted by full circles. Taken from Ref. [3].

PK1 [25, 26]. For comparison we present the results with and without the non-axial deformations. The reflection asymmetric shapes are allowed in both calculations. It was known that most of the outer barriers of the actinide nuclei are reflection asymmetric, which is consistent with the observed low-energy asymmetric fission fragments in this region. However, to estimate the fission barrier heights with an even higher accuracy, it is desirable to include also the effects of the non-axial deformations. From these potential energy curves it is clear that the non-axial deformations lower the outer barriers by around 0.5 to 1 MeV compared with the axial symmetric results. The empirical values of the fission barriers are also presented in the figure as full circles. Apparently, by including the triaxiality, the agreement between the theory and the experiment is improved.

In Ref. [6] the non-axial reflection-asymmetric β_{32} shape in some transfermium nuclei with $N = 150$, namely ^{246}Cm, ^{248}Cf, ^{250}Fm, and ^{252}No were investigated using the MDC-RMF model. In Fig. 5, we show potential energy curves for these $N = 150$ isotones. For ^{246}Cm, the ground state deformation $\beta_{32} = 0.020$. The potential energy curve is rather flat around the minimum. We denote the energy difference between the ground state and the point with $\beta_{32} = 0$ by $E_{\rm depth}$ which measures the energy gain with respect to the β_{32} distortion. For ^{246}Cm, $E_{\rm depth}$ is only 34 keV. For ^{248}Cf, ^{250}Fm, and ^{252}No, the minima locate at $\beta_{32} = 0.037$, 0.034, and 0.025, respectively. The corresponding energy gain $E_{\rm depth} = 0.351$, 0.328, and 0.104 MeV. As is discussed in Ref. [6], the occurrence of the non-axial octupole β_{32} correlations is mainly from

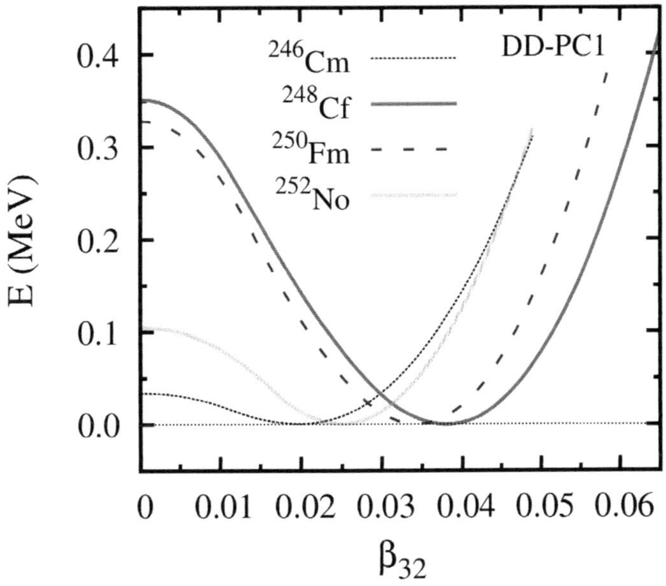

Figure 5. (Color online) The binding energy E (relative to the ground state) for $N = 150$ isotones ^{246}Cm (dashed line), ^{248}Cf (solid line), ^{250}Fm (long-dashed line), and ^{252}No (dotted line) as a function of the non-axial octupole deformation parameter β_{32}. Taken from Ref. [6].

a pair of neutron orbitals [734]9/2 ($\nu j_{15/2}$) and [622]5/2 ($\nu g_{9/2}$) which are close to the neutron Fermi surface and a pair of proton orbitals [521]3/2 ($\pi f_{7/2}$) and [633]7/2 ($\pi i_{13/2}$) which are close to the proton Fermi surface.

It may be hard to conclude that these nuclei have static non-axial octupole deformations from these results because the potential energy curve is flat around the minimum and E_{depth} is small [6]. However, the present calculations at least indicate a strong Y_{32}-correlation in these nuclei. Both the non-axial octupole parameter β_{32} and the energy gain E_{depth} reach maximal values at ^{248}Cf in the four nuclei along the $N = 150$ isotonic chain. This is consistent with the analysis given in Refs. [34, 35] and the experimental observation that in ^{248}Cf, the 2^- state is the lowest among these nuclei [36].

4. Summary

In this contribution we present the formalism and some applications of the multi-dimensional constrained relativistic mean field (MDC-RMF) models in which all shape degrees of freedom $\beta_{\lambda\mu}$ deformations with even μ are allowed. The potential energy surfaces (curves) of ^{240}Pu and actinide nuclei with various symmetries are investigated. It is found that besides the octupole deformation, the triaxiality also plays an important role upon the second fission barriers. For most of even-even actinide nuclei, the triaxiality lowers the outer barrier by $0.5 \sim 1$ MeV, accounting for about $10 \sim 20\%$ of the barrier height. The non-axial reflection-asymmetric β_{32} shape in some transfermium nuclei with $N = 150$, namely ^{246}Cm, ^{248}Cf, ^{250}Fm, and ^{252}No are studied and rather strong non-axial octupole Y_{32} effects have been found in ^{248}Cf and ^{250}Fm which are both well deformed with large axial-quadrupole deformations, $\beta_{20} \approx 0.3$. We note that it is crucial to include the reflection asymmetric and non-axial shapes simultaneously for the study of potential energy surfaces and fission barriers of actinide nuclei and of nuclei in unknown mass regions such as superheavy nuclei.

Acknowledgement
This work has been supported by Major State Basic Research Development Program of China (Grant No. 2013CB834400), National Natural Science Foundation of China (Grant Nos. 11121403, 11175252, 11120101005, 11211120152, and 11275248), Knowledge Innovation Project of Chinese Academy of Sciences (Grant Nos. KJCX2- EW-N01 and KJCX2-YW-N32). The results described in this paper are obtained on the ScGrid of Supercomputing Center, Computer Network Information Center of Chinese Academy of Sciences.

References

[1] Bohr A and Mottelson B R 1998 *Nuclear Structure* Vol I (World Scientific)
[2] Ring P and Schuck P 1980 *The Nuclear Many-Body Problem* (Springer-Verlag Berlin, Heidelberg, and New York)
[3] Lu B N 2012 *Multi-dimensional constrained relativistic mean field theory and the potential energy surfaces and fission barriers of actinide nuclei* Ph.D. thesis, Institute of Theoretical Physics, Chinese Academy of Sciences (in Chinese)
[4] Lu B N, Zhao J, Zhao E G and Zhou S G 2013 (*Preprint* 1304.2513)
[5] Lu B N, Zhao E G and Zhou S G 2012 *Phys. Rev. C* **85** 011301(R)–5 (*Preprint* 1110.6769)
[6] Zhao J, Lu B N, Zhao E G and Zhou S G 2012 *Phys. Rev. C* **86** 057304–4 (*Preprint* 1209.6567)
[7] Lu B N, Zhao J, Zhao E G and Zhou S G 2012 *EPJ Web of Conferences* **38** 05003, Proceedings of the Int. Conf. on Nuclear Structure and Related Topics (NSRT12), Dubna, Russia, 2-7 July 2012 (*Preprint* 1303.0621)
[8] Serot B D and Walecka J D 1986 *Adv. Nucl. Phys.* **16** 1–327
[9] Reinhard P G 1989 *Rep. Prog. Phys.* **52** 439–514
[10] Ring P 1996 *Prog. Part. Nucl. Phys.* **37** 193–263
[11] Vretenar D, Afanasjev A, Lalazissis G and Ring P 2005 *Phys. Rep.* **409** 101–259
[12] Meng J, Toki H, Zhou S G, Zhang S Q, Long W H and Geng L S 2006 *Prog. Part. Nucl. Phys.* **57** 470–563 (*Preprint* nucl-th/0508020)
[13] Paar N, Vretenar D and Colo G 2007 *Rep. Prog. Phys.* **70** 691–793
[14] Niksic T, Vretenar D and Ring P 2011 *Prog. Part. Nucl. Phys.* **66** 519–548
[15] Prassa V, Niksic T, Lalazissis G A and Vretenar D 2012 *Phys. Rev. C* **86** 024317–12
[16] Meng J, Peng J, Zhang S Q and Zhao P W 2013 *Frontiers of Physics* **8** 55–79
[17] Lu B N, Zhao E G and Zhou S G 2011 *Phys. Rev. C* **84** 014328–10 (*Preprint* 1104.4638)
[18] Gambhir Y K, Ring P and Thimet A 1990 *Ann. Phys.* **198** 132–179
[19] Ring P, Gambhir Y K and Lalazissis G A 1997 *Comput. Phys. Commun.* **105** 77–97
[20] Geng L S, Meng J and Toki H 2007 *Chin. Phys. Lett.* **24** 1865–1868
[21] Tian Y, Ma Z and Ring P 2009 *Phys. Lett. B* **676** 44–50
[22] Tian Y, Ma Z and Ring P 2009 *Phys. Rev. C* **80** 024313–7
[23] Niksic T, Ring P, Vretenar D, Tian Y and Ma Z 2010 *Phys. Rev. C* **81** 054318–12
[24] Capote R, Herman M, Oblozinsky P, Young P, Goriely S, Belgya T, Ignatyuk A, Koning A, Hilaire S, Plujko V, Avrigeanu M, Bersillon O, Chadwick M, Fukahori T, Ge Z, Han Y, Kailas S, Kopecky J, Maslov V, Reffo G, Sin M, Soukhovitskii E and Talou P 2009 *Nucl. Data Sheets* **110** 3107–3214
[25] Zhao P W, Li Z P, Yao J M and Meng J 2010 *Phys. Rev. C* **82** 054319–14
[26] Zhao P W, Song L S, Sun B, Geissel H and Meng J 2012 *Phys. Rev. C* **86** 064324–6
[27] Abusara H, Afanasjev A V and Ring P 2010 *Phys. Rev. C* **82** 044303–11
[28] Afanasjev A V 2013 Proceedings of Fifth International Conference on Fission and Properties of Neutron-rich Nuclei, November 4-10, 2012, Sanibel Island, USA (*Preprint* 1303.1206)
[29] Lalazissis G A, Konig J and Ring P 1997 *Phys. Rev. C* **55** 540–543
[30] Lalazissis G A, Karatzikos S, Fossion R, Arteaga D P, Afanasjev A V and Ring P 2009 *Phys. Lett. B* **671** 36–41
[31] Bender M, Rutz K, Reinhard P G, Maruhn J A and Greiner W 1999 *Phys. Rev. C* **60** 034304–20
[32] Lalazissis G A, Niksic T, Vretenar D and Ring P 2005 *Phys. Rev. C* **71** 024312–10
[33] Niksic T, Vretenar D and Ring P 2008 *Phys. Rev. C* **78** 034318–19
[34] Chen Y S, Sun Y and Gao Z C 2008 *Phys. Rev. C* **77** 061305(R)–5
[35] Jolos R V, Malov L A, Shirikova N Y and Sushkov A V 2011 *J. Phys. G: Nucl. Part. Phys.* **38** 115103
[36] Robinson A P, Khoo T L, Ahmad I, Tandel S K, Kondev F G, Nakatsukasa T, Seweryniak D, Asai M, Back B B, Carpenter M P, Chowdhury P, Davids C N, Eeckhaudt S, Greene J P, Greenlees P T, Gros S, Heinz A, Herzberg R D, Janssens R V F, Jones G D, Lauritsen T, Lister C J, Peterson D, Qian J, Tandel U S, Wang X and Zhu S 2008 *Phys. Rev. C* **78** 034308–6

Journal of Physics: Conference Series **492** (2014) 012015 doi:10.1088/1742-6596/492/1/012015

Phenomenological effects of CP conserving Higgs bosons self couplings in SM×S(3)

E Barradas Guevara[1], O Félix-Beltrán[2], F González-Canales[2], and E Rodríguez-Jáuregui[3]

[1]Facultad de Cs. Físico Matemáticas, BUAP, Puebla, Pue. México, C.P. 72570.

[2]Facultad de Cs. de la Electrónica, BUAP, Puebla, Pue. México.

[3]Departamento de Física, UNISON, Hermosillo, Sonora, México.

E-mail: barradas@fcfm.buap.mx

Abstract. A detailed analysis of a minimal S_3-invariant extension of the Standard Model including an extended S_3-Higgs sector is performed. In this model, we study the trilinear Higgs couplings and its dependence on the details of the model, even when the lightest Higgs boson mass is taken to be a fixed parameter. We study quantitatively the trilinear Higgs couplings, and compare these couplings to the corresponding Standard Model trilinear Higgs coupling in some regions of the parameter space. A precise measurement of the trilinear Higgs self coupling will also make it possible to test this extended $S(3)$-Standard Model which has a different trilinear Higgs couplings as compared to the Standard Model. Finally, partial numerical results of the phenomenological Higgs effects are presented.

1. Introduction

In the Standard Model (SM) the electroweak gauge bosons and the fundamental matter particles acquire masses through the interaction with a scalar field. The Higgs boson is a essential part of the SM upon the spontaneous breaking of the electroweak symmetry. The minimal version of the SM contains one complex Higgs doublet, resulting in one physical neutral CP-even Higgs boson h_{SM} after electroweak symmetry breaking (EWSB). Although their existence is a basic piece of the theory and the Higgs potential is very simple and sufficient to describe a realistic model of mass generation, this may not be the final form of the theory [1]. The minimal supersymmetric extension of the SM (MSSM) contains a scalar Higgs sector corresponding to that of a Two-Higgs Doublet Model (2HDM). Models with Higgs doublets (and singlets) possess the important phenomenological property that

$\rho = M_W/(M_Z \cos\theta_W) = 1$ up to finite radiative corrections. In these days when it is to be seen if the new boson at a mass of 125 GeV, observed in experiment CMS at the LHC, corresponding to the scaler of the Standard Model (SM), is of ultimate importance to analyze the mechanism of spontaneous symmetry breaking with models that have more number of scalar bosons [2].

Each family of fermions enters independently in the SM, so that to understand the replication of generations and to reduce the number of free parameters, usually more symmetry is introduced in the theory. In this direction interesting work has been done with the addition of discrete symmetries to the SM (see for instance [3] and references therein, for a review on the subject). It is noticeable that many interesting features of masses and mixing of the SM can be understood using a minimal discrete group, namely the permutational group S_3 [4].

Prior to the introduction of the Higgs boson, the SM is chiral and invariant with respect to any permutation of the left and right quark and lepton fields. After the introduction of the Higgs boson in the theory, this field may be treated as an S_3 singlet H_S, but then, only one fermion in each family can acquire mass. To give mass to all fermions and, at the same time, preserve the S_3 flavour symmetry of the theory, an extended flavoured Higgs sector is required with three Higgs $SU(2)$ doublets, one in a singlet and the other two in a doublet irreducible representation of S_3 [5].

Up to now, the particle observed at the LHC is a particle in the physical spectrum of the Higgs boson of the SM . It is not known if there is one or many Higgs bosons. An indication of the presence of one Higgs boson or an extended Higgs sector, as the one proposed in the S_3-invariant extension of the Standard Model, could be found at the Large Hadron Collider [6].

Models with more than one Higgs doublet, with or without supersymmetry, have been studied extensively. For a review of supersymmetric and two Higgs-doublet models, see for instance [7]. Different aspects of several three and more Higgs doublets models have also been studied, with and without discrete symmetries (see for instance [8]). In particular, in refs. [9] it was shown that in two-Higgs doublet models, at tree level, the potential minimum that preserves electric charge and CP symmetries, when it exists is stable and is the global one. Many of these models are not concerned with the unsolved problem of family replication, though. There are also analysis of different aspects of the Higgs potential of various discrete flavour groups, see for instance [10].

We study the trilinear Higgs couplings and its dependence on the details of the model, with an extended S_3-Higgs sector, even when the lightest Higgs boson mass is taken to be a fixed parameter. We study quantitatively the trilinear Higgs couplings, and compare these couplings to the corresponding Standard Model trilinear Higgs coupling in some regions of the parameter space. This paper is organized as follows: in section 2 we present some remarks about the S_3 flavour symmetry and its Lagrangian; the Higgs potential for this model: the Higgs sector of the S_3 extended model, and the form of the Higgs mass matrix; in section 3 we focus on the trilinear self-couplings of neutral Higgs bosons; whereas the details of the numerical results are presented in sections 4; and finally in section 5 we present our conclusions.

2. The $S(3)$ extended Higgs doublet model

The Lagrangian $\mathcal{L}_\Phi$ of the Higgs sector is given by

$$\mathcal{L}_\Phi = [D_\mu H_S]^2 + [D_\mu H_1]^2 + [D_\mu H_2]^2 - V(H_1, H_2, H_S), \tag{1}$$

where D_μ is the usual covariant derivative. The scalar potential $V(H_1, H_2, H_S)$ is the most general Higgs potential invariant under $SU(3)_C \times SU(2)_L \times U(1)_Y \times S_3$. The analysis of the stability properties of the potential V is of great relevance to study the phenomenological implications of this model. In here we will concentrate on the analysis of the potential V following the lines of previous work done in the vacuum stability of multi-Higgs models [9]. In our case, the discrete flavour symmetry S_3 simplifies the analysis.

There are many different ways of writing the Higgs potential for this model, but for the purpose of this work the best basis is

$$H_1 = \begin{pmatrix} \phi_1 + i\phi_4 \\ \phi_7 + i\phi_{10} \end{pmatrix}, \; H_2 = \begin{pmatrix} \phi_2 + i\phi_5 \\ \phi_8 + i\phi_{11} \end{pmatrix}, \; H_S = \begin{pmatrix} \phi_3 + i\phi_6 \\ \phi_9 + i\phi_{12} \end{pmatrix}, \tag{2}$$

the numbering of the real scalar ϕ fields is chosen for convenience in writing the mass matrices for the scalar particles and the subscript S is the flavor index for the Higgs field singlet. Now a simple matter to write down the potential is

$$\begin{aligned} V = \; & \mu_1^2(x_1 + x_2) + \mu_0^2 x_3 + a x_3^2 + b(x_1 + x_2)x_3 + c(x_1 + x_2)^2 \\ & -4d x_7^2 + 2e[(x_1 - x_2)x_6 + 2x_4 x_5] + f\left(x_5^2 + x_6^2 + x_8^2 + x_9^2\right) \\ & +g\left[(x_1 - x_2)^2 + 4x_4^2\right] + 2h\left(x_5^2 + x_6^2 - x_8^2 - x_9^2\right). \end{aligned} \tag{3}$$

where the $\mu_{0,1}^2$ parameters have dimensions of mass squared, the real couplings $a, \cdots, h$ are dimensionless and the invariants x_i, the potential V depends on the fields ϕ_i through x_i, considering our assignment as

$$\begin{aligned} x_1 &= H_1^\dagger H_1, \quad x_4 = \mathcal{R}\left(H_1^\dagger H_2\right), \quad x_7 = \mathcal{I}\left(H_1^\dagger H_2\right), \\ x_2 &= H_2^\dagger H_2, \quad x_5 = \mathcal{R}\left(H_1^\dagger H_S\right), \quad x_8 = \mathcal{I}\left(H_1^\dagger H_S\right), \\ x_3 &= H_S^\dagger H_S, \quad x_6 = \mathcal{R}\left(H_2^\dagger H_S\right), \quad x_9 = \mathcal{I}\left(H_2^\dagger H_S\right). \end{aligned} \tag{4}$$

If the Higgs potential S_3 invariant Eq.(3) is bounded from below, being a quartic polynomial function it will certainly have a global minimum somewhere. We can two types of minima: the "trivial" one, for which the Higgs acquires zero VEV's, and the usual one, where electroweak symmetry breaking occurs, away from the origin, for

$$\phi_7 = v_1,\ \phi_8 = v_2,\ \phi_9 = v_3,\ \phi_i = 0,\ \ i \neq 7, 8, 9, \tag{5}$$

defined as the normal minimum, with VEV's which do not have any complex relative phase. Where, v_1, v_2 and v_3 are the VEV's of the unbroken Higgs fields.

The minimization constraints may then be determined by demanding the vanishing of $\partial V/\partial\phi_i$, then we get the mass parameter by

$$\mu_1^2 = -(b+f+2h)\,v_3^2 - 2\,(c+g)\left(v_1^2+v_2^2\right) + \frac{3e(v_1^2-2v_1v_2-v_2^2)v_3}{v_1-v_2}, \tag{6}$$

and

$$\mu_0^2 = -\left[2av_3^2 + (b+f+2h)\left(v_1^2+v_2^2\right) - e\left(\frac{3v_1^2-v_2^2}{v_3}\right)v_2\right]. \tag{7}$$

As well as, we have $v_1 = \sqrt{3}v_2$.

The Higgs-bosons masses in this model are obtained by diagonalizing the 12×12 matrix, with $M_I^2 = M_{II}^2$ and M_{IJ}^2, $I, J = C,\ S,\ P$, are all 3×3 sub-matrices. From

$$(\mathcal{M}_H^2)_{ij} = \frac{1}{2}\frac{\partial^2 V}{\partial\phi_i\partial\phi_j}\bigg|_{\text{mim}}, \tag{8}$$

with $i, j = \overline{1, 12}$. We have

$$\mathcal{M}_H^2 = \text{diag}\left(M_C^2, M_C^2, M_S^2, M_P^2\right). \tag{9}$$

The matrix charged Higgs is two times degenerate,

$$M_C^2 = \begin{pmatrix} c_{11} & c_{12} & c_{13} \\ & c_{22} & c_{23} \\ & & c_{33} \end{pmatrix}, \tag{10}$$

where

$$\begin{aligned}
c_{11} &= -4gv_2^2 - v_3(4ev_2 + (f+2h)v_3) \\
c_{12} &= 2\sqrt{3}v_2(2gv_2 + ev_3) \\
c_{13} &= \sqrt{3}v_2\,(2ev_2 + (f+2h)v_3) \\
c_{22} &= -12gv_2^2 - v_3\,(8ev_2 + (f+2h)v_3) \\
c_{23} &= v_2\,(2ev_2 + (f+2h)v_3) \\
c_{33} &= -4v_2^2\,(2ev_2 + (f+2h)v_3)\,/v_3.
\end{aligned}$$

The mass matrix for the CP-even Higgs scalars,

$$M_S^2 = \begin{pmatrix} s_{11} & s_{12} & s_{13} \\ & s_{22} & s_{23} \\ & & s_{33} \end{pmatrix}, \tag{11}$$

where

$$\begin{aligned}
s_{11} &= 12(c+g)v_2^2 \\
s_{12} &= 2\sqrt{3}v_2\left(2(c+g)v_2 + 3ev_3\right) \\
s_{13} &= 2\sqrt{3}v_2\left(3ev_2 + (b+f+2h)v_3\right) \\
s_{22} &= 4v_2\left((c+g)v_2 - 3ev_3\right) \\
s_{23} &= 2v_2\left(3ev_2 + (b+f+2h)v_3\right) \\
s_{33} &= 4av_3^2 - 8ev_2^3/v_3.
\end{aligned}$$

The mass matrix for the CP-odd Higgs pseudoscalar,

$$M_P^2 = \begin{pmatrix} p_{11} & p_{12} & p_{13} \\ & p_{22} & p_{23} \\ & & p_{33} \end{pmatrix}, \tag{12}$$

where

$$\begin{aligned}
p_{11} &= -4\left((d+g)v_2^2 + ev_3v_2 + hv_3^2\right) \\
p_{12} &= 2\sqrt{3}v_2\left(2(d+g)v_2 + ev_3\right) \\
p_{13} &= 2\sqrt{3}v_2(ev_2 + 2hv_3) \\
p_{22} &= -12(d+g)v_2^2 - 8ev_2v_3 - 4hv_3^2 \\
p_{23} &= 2v_2(ev_2 + 2hv_3) \\
p_{33} &= -8v_2^2(ev_2 + 2hv_3)/v_3.
\end{aligned}$$

After diagonalizing the mass matrices the masses of the physical scalars/pseudoscalars are obtained. In our analysis we are not taking into account the parameter space with negative eigenvalues solutions for the squared masses of the physical Higgs fields. Of the original twelve scalar degrees of freedom, three Goldstone bosons ($G^\pm$ and G) are absorbed by $W^\pm$ and Z. The remaining nine physical Higgs particles are three $CP-$even scalar (h and H_1, H_2, with $m_h \leq m_{H_1} \leq m_{H_2}$), two CP-odd scalar (A_1, and A_2, with $m_{A_1} \leq m_{A_2}$), and two charged Higgs pair ($H^\pm_{1,2}$, mass degenerate). Their physical masses and the respective Higgs-scalar mixing are related to the weak parameters of the Higgs potential as follows. We stars considering the mass matrix for the CP-even Higgs scalars Eq.(11).

Defining the physical mass eigenstates m_h^2, $m_{H_1}^2$, and $m_{H_2}^2$, the masses are found from the diagonalization process

$$\mathcal{M}^2_{diag} = R^T \mathcal{M}^2_{even} R = \text{diag}(m_h^2, m_{H_1}^2, m_{H_2}^2). \tag{13}$$

The masses for the CP-even Higgs scalars are:

$$m_h^2 = -18ev_2v_3$$

$$m^2_{H_1^0,H_2^0} = (\mathcal{M}_a^2 + \mathcal{M}_c^2) \pm \sqrt{(\mathcal{M}_a^2 - \mathcal{M}_c^2)^2 + (\mathcal{M}_b^2)^2}, \tag{14}$$

where

$$\begin{aligned}\mathcal{M}_a^2 &= v_2(8(c+g)v_2 + 3ev_3)\\ \mathcal{M}_b^2 &= 4v_2\,(3ev_2 + (b+f+2h)v_3)\\ \mathcal{M}_c^2 &= -\frac{4ev_2^3}{v_3} + 2av_3^2,\end{aligned} \tag{15}$$

and mixing angles are defined as:

$$\tan 2\alpha_3 = \frac{\mathcal{M}_b^2}{\mathcal{M}_a^2 - \mathcal{M}_c^2} \tag{16}$$

As well as,

$$R = O \cdot P \cdot Q, \tag{17}$$

$$O = \begin{pmatrix} -1 & 0 & 0 \\ 0 & 1 & 0 \\ 0 & 0 & -1 \end{pmatrix}, \quad P = \begin{pmatrix} 1/2 & -\sqrt{3}/2 & 0 \\ \sqrt{3}/2 & 1/2 & 0 \\ 0 & 0 & 1 \end{pmatrix}, \quad Q = \begin{pmatrix} 1 & 0 & 0 \\ 0 & c_\theta & s_\theta \\ 0 & -s_\theta & c_\theta \end{pmatrix}, \tag{18}$$

where

$$\tan\theta = \frac{s_\theta}{c_\theta} = \frac{2s_{13}}{\sqrt{3}(m_{H_1}^2 - s_{33})} = \frac{\mathcal{M}_b^2}{2\mathcal{M}_a^2 - m_{H_2}^2} \tag{19}$$

which s_{13} and s_{33} from Eq.(11).

To generate the correct $W^\pm$ and Z^0 masses, with the assignments $v_1^2 + v_2^2 + v_3^2 = v^2$ has to hold, where $v = 246$ GeV, and $v_1 = \sqrt{3}v_2$. Also, the v_i are taken real, i.e., we assume that spontaneous CP violation does not occur. The squared-mass parameters μ_0^2 and μ_1^2 can be eliminated by minimizing the scalar potential. The resulting squared masses for the CP-odd Higgs pseudoscalar are

$$m_{A_1}^2 = -16(d+g)v_2^2 - 10ev_2v_3 - 4hv_3^2, \qquad m_{A_2}^2 = -\frac{2(ev_2 + 2hv_3)(4v_2^2 + v_3^2)}{v_3}.$$

can be parametrized with

$$\tan\omega_3 = \frac{2v_2}{v_3} \tag{20}$$

where $s\omega_3 = 2v_2/v$, $c\omega_3 = v_3/v$.

The charged Higgs boson masses are

$$m_{H_1^\pm}^2 = -\left(10ev_2 + (f+2h)v_3\right)v_3, \qquad m_{H_2^\pm}^2 = -\frac{v^2}{v_3}\left(2ev_2 + (f+2h)v_3\right).$$

Higgs boson masses are not determined *a priori* within the theory and their decay patterns depend strongly on the masses, in order to determine the possible decay modes and branching ratios, it is necessary to investigate the change of mass spectrum with respect to the quadrilinears $a, \ldots, h$ couplings. For simplicity, we assume in our numerical analysis for the following dimensionless parameters

$$a = 1, \quad b = 1, \quad c = 1, \quad d = -1, \quad e = -5/3, \quad f = 3.25, \quad g = 1, \quad h = 1,$$

and $v_2 = 246(\sin\omega_3)/2$ GeV, $v_3 = 246\cos\omega_3$ GeV, $-\pi \le \omega_3 \le \pi$. The Figure 1 shows the masses of three CP-even Higgs scalars with respect to ω_3.

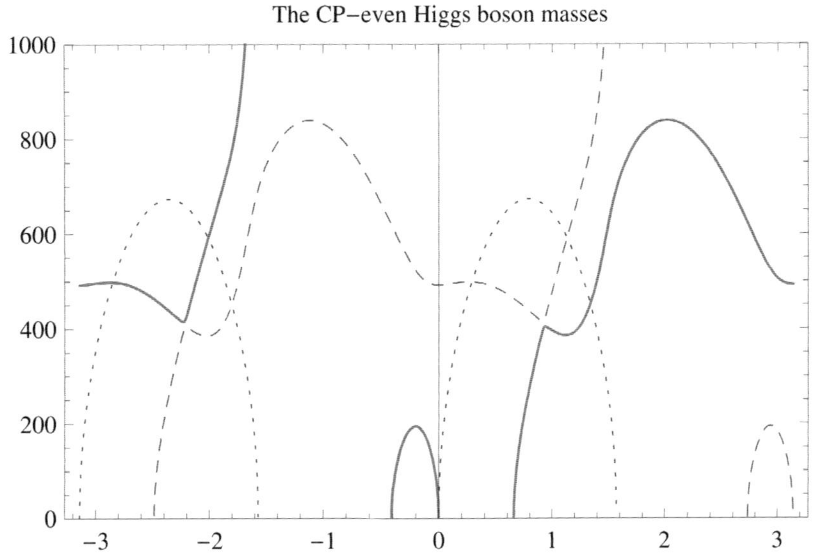

Figure 1. The masses of three CP-even Higgs scalars with respect to ω_3. The solid line is for H_1^0, the dashed line is for H_2^0 and the dotted line is for h^0.

The quadrilinears $a, \ldots, h$ couplings are related to the masses of the Higgs bosons by

$$a = \frac{9\mathcal{M}_c^2 v_3^2 - 2m_{h^0}^2 v_2^2}{18 v_3^4}$$

$$b = \frac{2m_{h^0}^2 v2 + 9\mathcal{M}_b^2 v_3}{36 v_2 v_3^2} + \frac{m_{H_2^\pm}^2}{v^2} \tag{21}$$

$$c = \frac{m_{H_2^\pm}^2}{4v^2} - \frac{m_{h^0}^2 - 9(m_{H_1^\pm}^2 - m_{H_2^\pm}^2 + 2\mathcal{M}_a^2)}{144 v_2^2}$$

where

$$\mathcal{M}_a^2 = \frac{1}{4}\left(m^2_{H_1^0} + m^2_{H_2^0} - (m^2_{H_1^0} - m^2_{H_2^0})\cos 2\alpha_3\right)$$

$$\mathcal{M}_b^2 = (\mathcal{M}_a^2 - \mathcal{M}_c^2)\tan 2\alpha_3 \tag{22}$$

$$\mathcal{M}_c^2 = \frac{1}{4}\left(m^2_{H_1^0} + m^2_{H_2^0} + (m^2_{H_1^0} - m^2_{H_2^0})\cos 2\alpha_3\right).$$

As well as,

$$d = \frac{1}{16}\left(\frac{m^2_{A_2} - m^2_{A_1} + m^2_{H_1^\pm} - m^2_{H_2^\pm}}{v_2^2} - \frac{m^2_{A_2} - m^2_{H_2^\pm}}{4v^2}\right)$$

$$e = -\frac{m^2_{h^0}}{(18 v_2 v_3)}$$

$$f = \frac{1}{18}\left(\frac{9(m^2_{A_2} - 2m^2_{H_2^\pm})}{v^2} + \frac{m^2_{h^0}}{v_3^2}\right) \tag{23}$$

$$g = \frac{4m^2_{h^0} - 9(m^2_{H_1^\pm} - m^2_{H_2^\pm})}{144 v_2^2} - \frac{m^2_{H_2^\pm}}{4v^2}$$

$$h = \frac{1}{36}\left(\frac{m^2_{h^0}}{v_3^2} - \frac{9m^2_{A_2}}{v^2}\right).$$

In Figure 2 shows the behavior of the links quartic function of the reason for expectation values v_2 and v_3 given by ω_3, Eq.(20). You can specify a value of ω_3 and from this, determine a mass spectrum of Higgs bosons. For example, $\omega_3 = 1$, quadrilinears couplings take values:

$$a = 3.58, \quad b = 5.60, \quad c = 3.40,$$

$$d = 1.36, \quad e = -0.06, \quad f = -3.57, \quad g = -2.39, \quad h = -0.23,$$

fixing values mass spectrum:

$$m_{h_0} = 125 \text{ GeV}, \quad m_{H_1} = 550 \text{ GeV}, \quad m_{H_2} = 336 \text{ GeV}.$$

$$m_{H_1^\pm} = 687 \text{ GeV}, \quad m_{H_2^\pm} = 488 \text{ GeV}.$$

$$m_{A_1^0} = 430 \text{ GeV}, \quad m_{A_2^0} = 225 \text{ GeV}.$$

3. Trilinear Self-Couplings of Neutral Higgs Bosons

The measurement of the Higgs self-coupling is crucial to determine the Higgs potential. The self-couplings are uniquely determined in the SM by the mass of the Higgs boson which is

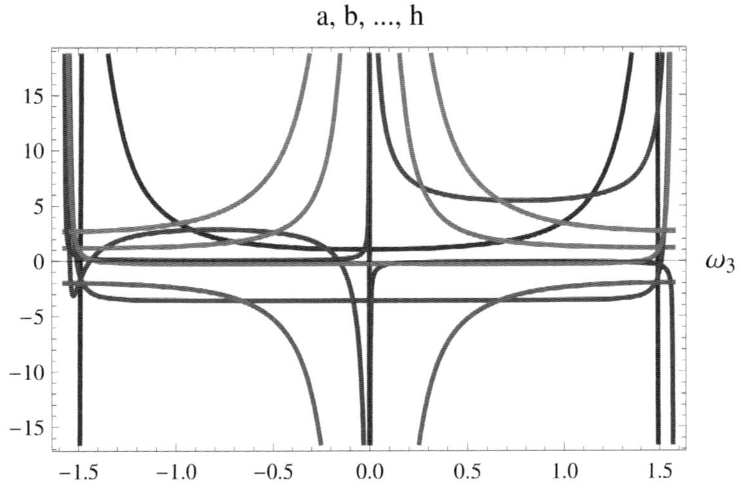

Figure 2. The quadrilinears $a, \ldots, h$ couplings with respect to ω_3, in the range $-\pi/2 \leq \omega_3 \leq \pi/2$. The ω_3 values close to zero are on the order of tenths, $\tan\omega_3 \sim 0.09$. When ω_3 approaches to $\pi/2$ diverges $\tan\omega_3$. Higher values are acceptable for $\tan\omega_3 = 5$. If $\tan\omega_3 = 1$, and then the mass spectrum of bosons Higgs is: $m_{h_0} = 125$ GeV, $m_{H_1} = 550$ GeV, $m_{H_2} = 315$ GeV, $m_{A_1^0} = 492$ GeV, $m_{A_2^0} = 232$ GeV, $m_{H_1^\pm} = 756$ GeV, $m_{H_2^\pm} = 491$ GeV. But if $\tan\omega_3 = 5$ as $\omega_3 = 1.3734$, then the mass spectrum of bosons Higgs is: $m_{h_0} = 125$ GeV, $m_{H_1} = 550$ GeV, $m_{H_2} = 305$ GeV, $m_{A_1^0} = 356$ GeV, $m_{A_2^0} = 57$ GeV, $m_{H_1^\pm} = 623$ GeV, $m_{H_2^\pm} = 437$ GeV

related to the quadrilinear coupling λ by $M_H = \sqrt{2\lambda}v$. The trilinear and quadrilinear vertices of the Higgs field H are given by the coefficients:

$$\lambda_{HHH} = \lambda v = \frac{M_H^2}{2v}, \qquad \lambda_{HHHH} = \frac{\lambda}{4} = \frac{M_H^2}{8v^2}. \tag{24}$$

The following definitions are often used:

$$\lambda_{ijk} = \frac{-i\partial^3 V}{\partial H_i \partial H_j \partial H_k}, \tag{25}$$

which are most easily obtained from the corresponding derivatives of V in Eq.(3) with respect to the fields $\{\phi_i\}$ with $i = 1, \cdots, 12$. We can then write the trilinear couplings in terms of the derivatives of the potential (3) with respect to ϕ_i and the elements of the rotation matrix R Eq.(17) as

$$\lambda_{ijk} = N \sum_{lmn} R_{il} R_{jm} R_{kn} \frac{\partial^3 V}{\partial\phi_i \partial\phi_j \partial\phi_k}, \tag{26}$$

where the indices l, m, n refer to the weak field basis, and $l \leq m \leq n = 1, 2, 3$, N is a factor of $n!$ for n identical fields. We now proceed to obtain these couplings in an explicit form.

The trilinear self-couplings a_{lmn} among the neutral Higgs bosons even can be written as

$$\begin{aligned} a_{1,1,1} &= 6\sqrt{6}(c+g)v_2, & a_{1,1,2,} &= \sqrt{2}(2(c+g)v_2 + 3ev_3), \\ a_{1,1,3} &= \sqrt{2}(3ev_2 + (b+f+2h)v_3), & a_{1,2,2} &= 2\sqrt{6}(c+g)v_2, \\ a_{1,2,3} &= 3\sqrt{6}ev_2, & a_{1,3,3} &= \sqrt{6}(b+f+2h)v_2, \\ a_{2,2,2} &= 3\sqrt{2}(2(c+g)v_2 - ev_3), & a_{2,2,3} &= \sqrt{2}((b+f+2h)v_3 - 3ev_2), \\ a_{2,3,3} &= \sqrt{2}(b+f+2h)v_2, & a_{3,3,3} &= 6\sqrt{2}av_3. \end{aligned} \tag{27}$$

Then, we here have self-couplings to three Higgs bosons even substituting for the elements of the rotation matrix, Eq.(17), one obtains

$$\begin{aligned} \lambda_{1,1,1} &= 6v\left(\lambda_1 s\omega_3 + \lambda_2 c\omega_3\right), & \lambda_{2,2,2} &= 6v\left(\lambda_3 s\omega_3 + \lambda_4 c\omega_3\right), \\ \lambda_{3,3,3} &= 6v\left(\lambda_5 s\omega_3 + \lambda_6 c\omega_3\right), & \lambda_{1,1,2} &= 2v\left(\lambda_7 s\omega_3 + \lambda_8 c\omega_3\right), \\ \lambda_{1,1,3} &= 2v\left(\lambda_9 s\omega_3 + \lambda_{10} c\omega_3\right), & \lambda_{1,2,2} &= 2v\left(\lambda_{11} s\omega_3 + \lambda_{12} c\omega_3\right), \\ \lambda_{1,2,3} &= v\left(\lambda_{13} s\omega_3 + \lambda_{14} c\omega_3\right), & \lambda_{1,3,3} &= 2v\left(\lambda_{15} s\omega_3 + \lambda_{16} c\omega_3\right), \\ \lambda_{2,2,3} &= 2v\left(\lambda_{17} s\omega_3 + \lambda_{18} c\omega_3\right), & \lambda_{2,3,3} &= 2v\left(\lambda_{19} s\omega_3 + \lambda_{20} c\omega_3\right). \end{aligned} \tag{28}$$

Where $\lambda_1, \ldots, \lambda_{20}$ depend to parameters quadrilinear couplings of the Higgs potential Eq.(3) and the mixing angle θ Eq.(19). For example, one can use the results of appendix to compute the first $\lambda_{h_0h_0h_0} = \lambda_{1,1,1}$, these results are given en the appendix,

$$\begin{aligned} \lambda_{h_0h_0h_0} &= \frac{\sqrt{3}}{4}v\left[2\left(18as_\theta^3 + (b+f+2h)(3c_\theta^2+1)s_\theta - 9ec_\theta^3 + 3ec_\theta\right)c\omega_3 \right. \\ &+ \left(3(b+f+2h)\left(c_\theta - 1\right)s_\theta^2 + 2(c+g)(9c_\theta^3 - 3c_\theta^2 + c_\theta - 3)\right. \\ &\left.\left. -3e(3c_\theta(c_\theta+1)-1)s_\theta\right)s\omega_3\right]. \end{aligned} \tag{29}$$

4. Numerical results

Here we present numerical results considering that is one a CP-even boson Higgs scalar behaves like the Higgs boson of the standard model, where the mass of the lightest Higgs scalar is significantly smaller than the masses of the other Higgs bosons of the model, We assume it is light and its mass is around 125 GeV. In our analysis we propose to H_1^0 as our candidate, it is a parameter free. The other bosons Higgs have masses consequently the order $\mathcal{O}(500)$ GeV and without loss of generality and for the sake of simplicity, consider that they have no equal masses. The Figure 3 shows the trilinear couplings by CP-even Higgs Bosons in function the mixing angle θ, Eq.(19), where $\tan\omega_3 = 1$. The coupling of higher intensity corresponds to $\lambda_{h_0h_0h_0}$ and $\lambda_{H_2H_2H_2}$, whereas that $\lambda_{H_1H_1H_1}$ to be very small,

$$\lambda_{h_0h_0h_0} \sim \lambda_{H_2H_2H_2} > \lambda_{H_1H_1H_1},$$

followed by $\lambda_{h_0h_0H_2}$, $\lambda_{h_0h_0H_1}$ $(\lambda_{h_0H_2H_2})$ and then the other. We proposed that the lightest is H_1^0 and its trilinear coupling is the lower intensity. Although there are couplings equal in

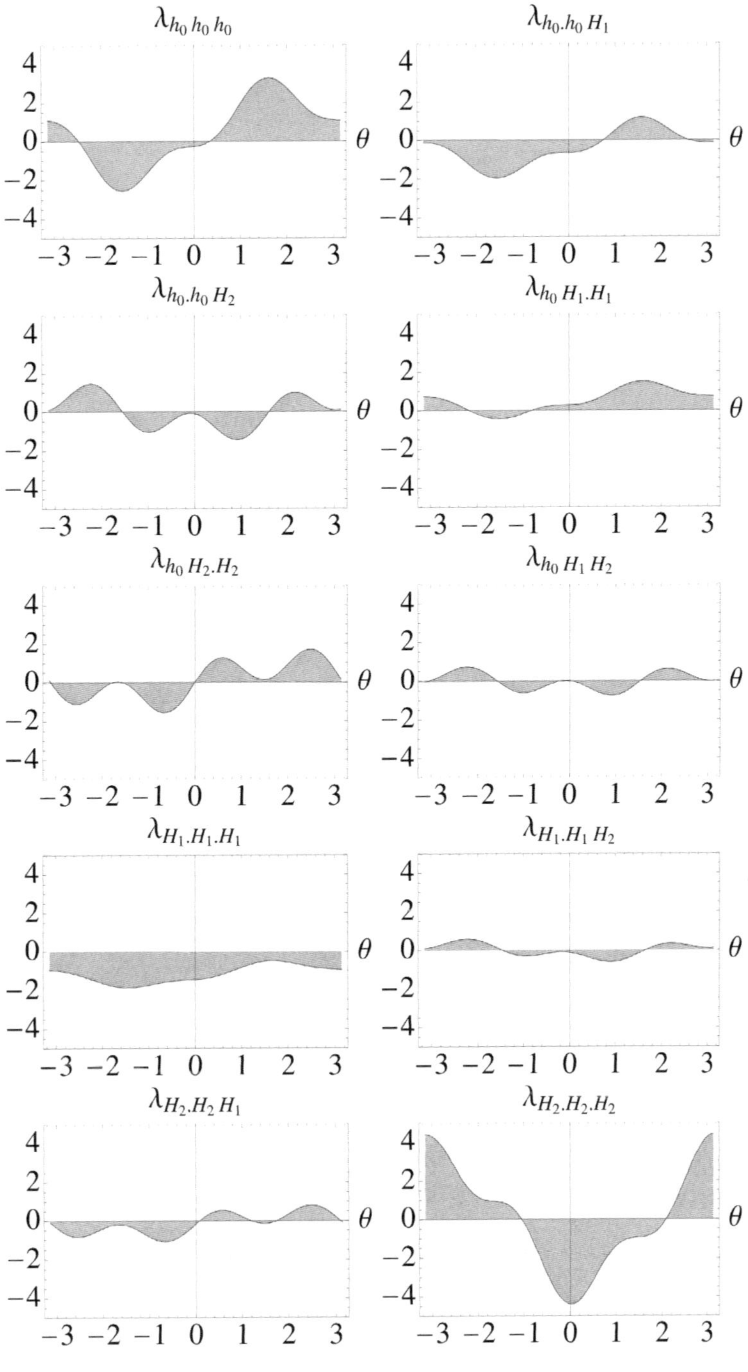

Figure 3. The trilinear couplings by CP-even Higgs bosons, $\lambda_{ijk}/(125\ \text{GeV})^2$ with $i,\ j,\ k = h_0,\ H_{1,2}$, in function the mixing angle $-\pi \leq \theta \leq \pi$, Eq.(19), where $\tan\omega_3 = 1$.

shape, they are different in intensity. Comparing these results with the SM trilinear coupling only. We see that our values are within acceptance ranges in the literature, but they are richer for a Higgs potential analysis, enabling better understand the spontaneous symmetry breaking weak.

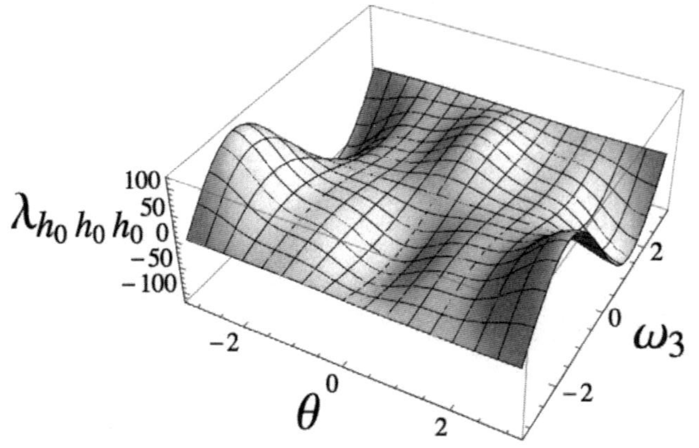

Figure 4. The trilinear couplings by CP-even Higgs bosons, $\lambda_{ijk}/(125\ \text{GeV})^2$ with $i,\ j,\ k = h_0,\ H_{1,2}$, in function the mixing angle $-\pi \leq \theta \leq \pi$, and $-\pi \leq \omega_3 \leq \pi$.

Furthermore, when comparing the self couplings of the Higgs boson as a function of the two mixing angles θ and ω_3, there is a symmetry, Figure 4. This allows us caracteizar different Higgs bosons. The mass of the Higgs boson is fixed through ω_3 and couplings are determined also by mixing Higgs bosons depending on θ. Our results are normalized $M_h = 125$ GeV in Figure 3, but with either parameterized three scalar bosons dependence is observed. See figure such. We can analyze the behavior of the car couplings, depending on the mass of the boson M_{H_2} through ω_3, see Figure 5.

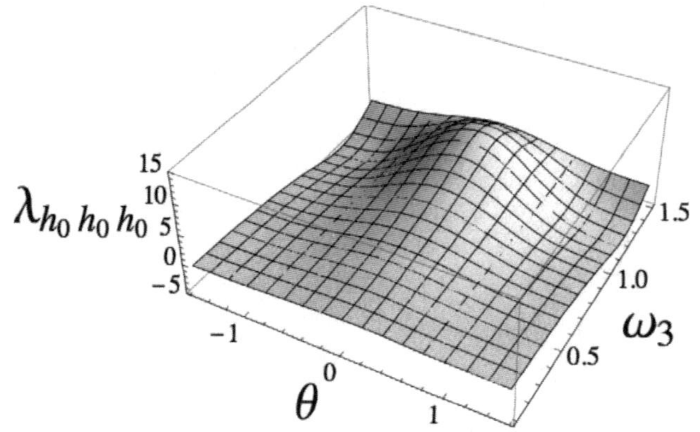

Figure 5. The trilinear couplings by CP-even Higgs bosons, $\lambda_{ijk}/(m_{H_2}\ \text{GeV})^2$ with $i,\ j,\ k = h_0,\ H_{1,2}$, in function the mixing angle $-\pi/2 \leq \theta \leq \pi/2$, and $0.1 \leq \omega_3 \leq 1.5$.

Journal of Physics: Conference Series **492** (2014) 012015 doi:10.1088/1742-6596/492/1/012015

5. Conclusions

We studied only the scalar sector assuming the pseudoscalars to be too heavy to be relevant. In this work we have analyzed the complete scalar sector of an S_3 flavour model. We deal with three CP-even, two CP-odd and two sets of charged scalar particles. We have improved our potential minimization technique which enabled us to explore a larger region of the allowed parameter space. We have studied in detail the trilinear couplings of the lightest Higgs boson of this model. Within the allowed domain of the parameters space of the model, the trilinear Higgs couplings have a strong dependence on $\tan\omega_3 = 2v_2/v_3$ and $\tan\theta$. The extended Higgs spectrum in S_3 models gives rise numbers of trilinear couplings. The hhh coupling can be measured in hh continuum production linear colliders as at e^+e^-.

Appendix

$$\begin{aligned}
\lambda_{h_0h_0h_0} &= 6v\left(\lambda_1 s\omega_3 + \lambda_2 c\omega_3\right), \\
\lambda_1 &= \frac{3\sqrt{3}}{4}(b+f+2h)(c_\theta - 1)s_\theta^2 \\
\lambda_2 &= \frac{2\sqrt{3}}{4}\left(18as_\theta^3 + \left[(b+f+2h)s_\theta - 3ec_\theta\right](3c_\theta^2+1)\right) \\
&\quad +2(c+g)(9c_\theta^3 - 3c_\theta^2 + c_\theta - 3) - 3e(3c_\theta(c_\theta+1)-1)s_\theta
\end{aligned} \tag{30}$$

$$\begin{aligned}
\lambda_{H_1H_1H_1} &= 6v\left(\lambda_3 s\omega_3 + \lambda_4 c\omega_3\right), \\
\lambda_3 &= \frac{1}{4}\left((b+f+2h)(c_\theta+3)s_\theta^2 + 6(c+g)(c_\theta^3 + c_\theta^2 + c_\theta + 9)\right. \\
&\quad \left. -3e((c_\theta-3)c_\theta - 3)s_\theta\right) \\
\lambda_4 &= \frac{1}{2}\left(6as_\theta^3 + (b+f+2h)(c_\theta^2+3)s_\theta - 3ec_\theta(c_\theta^2-3)\right)
\end{aligned} \tag{31}$$

$$\begin{aligned}
\lambda_{H_2H_2H_2} &= 6v\left(\lambda_5 s\omega_3 + \lambda_6 c\omega_3\right), \\
\lambda_5 &= 2\left((b+f+2h)c_\theta^2 + 6(c+g)s_\theta^2 + 3ec_\theta s_\theta\right) \\
\lambda_6 &= -4\left(6ac_\theta^3 + s_\theta^2((b+f+2h)c_\theta + 3es_\theta)\right)
\end{aligned} \tag{32}$$

$$\begin{aligned}
\lambda_{h_0h_0H_1} &= 2v\left(\lambda_7 s\omega_3 + \lambda_8 c\omega_3\right), \\
\lambda_7 &= \frac{1}{4}\left(3(b+f+2h)(3c_\theta+1)s_\theta^2\right. \\
&\quad \left. +2(c+g)(c_\theta(3c_\theta(9c_\theta+1)-5)+27) - 3e(9c_\theta^2 - 3c_\theta + 5)s_\theta\right) \\
\lambda_8 &= \frac{1}{4}\left(54as_\theta^3 + (b+f+2h)(9c_\theta^2-5)s_\theta - 3ec_\theta(9c_\theta^2+5)\right)
\end{aligned} \tag{33}$$

$$\begin{aligned}
\lambda_{h_0h_0H_2} &= 2v\left(\lambda_9 s\omega_3 + \lambda_{10} c\omega_3\right), \\
\lambda_9 &= \frac{1}{2}\left(2s_\theta(3(b(1-c_\theta) + (9c-f+9g-2h)c_\theta - 2c + f - 2g + 2h)c_\theta + g)\right. \\
&\quad \left. +3(b+f+2h)s_\theta^3 + 2cs_\theta - 9e(2c_\theta+1)s_\theta^2 + 3ec_\theta(3c_\theta(c_\theta+1)-1)\right) \\
\lambda_{10} &= 6(-9a+b+f+2h)c_\theta s_\theta^2 - (b+f+2h)(3c_\theta^3 + c_\theta) - 3e(9c_\theta^2-1)s_\theta
\end{aligned} \tag{34}$$

$$\begin{aligned}
\lambda_{h_0 H_1 H_1} &= 2v\left(\lambda_{11} s\omega_3 + \lambda_{12} c\omega_3\right), \\
\lambda_{11} &= \frac{\sqrt{3}}{4}\left((b+f+2h)(3c_\theta+5)s_\theta^2 + 2(c+g)(9c_\theta^3+5c_\theta^2+c_\theta-27)\right. \\
&\quad \left. +3e((5-3c_\theta)c_\theta+1)s_\theta\right) \\
\lambda_{12} &= \frac{\sqrt{3}}{2}\left(18as_\theta^3 + (b+f+2h)(3c_\theta^2+1)s_\theta - 9ec_\theta^3 + 3ec_\theta\right)
\end{aligned} \tag{35}$$

$$\begin{aligned}
\lambda_{h_0 H_1 H_2} &= v\left(\lambda_{13} s\omega_3 + \lambda_{14} c\omega_3\right), \\
\lambda_{13} &= \sqrt{3}\left(-2((b(1+c_\theta) - 9cc_\theta - 2c + c_\theta f - 9c_\theta g + 2(c_\theta+1)h + f - 2g)c_\theta + g)s_\theta\right. \\
&\quad \left. +(b+f+2h)s_\theta^3 - 2cs_\theta + 3e(1-2c_\theta)s_\theta^2 + 3ec_\theta((c_\theta-1)c_\theta+1)\right) \\
\lambda_{14} &= 2\sqrt{3}\left((2(-9a+b+f+2h)s_\theta^2 - (b+f+2h)(c_\theta^2-1) - 9ec_\theta s_\theta)c_\theta - 3es_\theta\right)
\end{aligned} \tag{36}$$

$$\begin{aligned}
\lambda_{h_0 H_2 H_2} &= 2v\left(\lambda_{15} s\omega_3 + \lambda_{16} c\omega_3\right), \\
\lambda_{15} &= \sqrt{3}\left(-2((b+f-9g+2h)c_\theta - 9cc_\theta + c + g)s_\theta^2 + (b+f+2h)(c_\theta-1)c_\theta^2\right. \\
&\quad \left. +3ec_\theta(2c_\theta+1)s_\theta - 3es_\theta^3\right) \\
\lambda_{16} &= 2\sqrt{3}s_\theta\left(18ac_\theta^2 - (b+f+2h)(2c_\theta^2 - s_\theta^2) - 9ec_\theta s_\theta\right)
\end{aligned} \tag{37}$$

$$\begin{aligned}
\lambda_{H_1 H_1 H_2} &= 2v\left(\lambda_{17} s\omega_3 + \lambda_{18} c\omega_3\right), \\
\lambda_{17} &= \frac{1}{2}\left(-2((b(c_\theta+3) + (-9c+f-9g+2h)c_\theta - 6c + 3f - 6g + 6h)c_\theta - 3g)s_\theta\right. \\
&\quad \left. +(b+f+2h)s_\theta^3 + 6cs_\theta + 3e(3-2c_\theta)s_\theta^2 + 3ec_\theta((c_\theta-3)c_\theta-3)\right) \\
\lambda_{18} &= 2(-9a+b+f+2h)c_\theta s_\theta^2 - (b+f+2h)(c_\theta^2+3)c_\theta - 9e(c_\theta^2-1)s_\theta
\end{aligned} \tag{38}$$

$$\begin{aligned}
\lambda_{H_1 H_2 H_2} &= 2v\left(\lambda_{19} s\omega_3 + \lambda_{20} c\omega_3\right), \\
\lambda_{19} &= 2(-(b+f-9g+2h)c_\theta + c(9c_\theta+3) + 3g)s_\theta^2 \\
&\quad +(b+f+2h)(c_\theta+3)c_\theta^2 + 3ec_\theta(2c_\theta-3)s_\theta - 3es_\theta^3 \\
\lambda_{20} &= 2\left(18ac_\theta^2 - (b+f+2h)(2c_\theta^2 - s_\theta^2) - 9ec_\theta s_\theta\right)
\end{aligned} \tag{39}$$

References

[1] Higgs P W 1964 Phys. Lett. **12** 132; Higgs P W 1966 Phys. Rev. **145** 1156; Englert F and Brout R 1964 Phys. Rev. Lett. **13** 321; Guralnik G S, Hagen C R and Kibble T W 1964 Phys. Rev. Lett. **13** 585.

[2] CMS Collaboration 2012 Physics Letters B 716 30-61.

[3] Ma E 2007 arXiv:0705.0327

[4] Derman E 1979 Phys. Rev. D **19**, 317; Derman E and Tsao H S 1979, Phys. Rev. D **20**, 1207; Mondragon A and Rodriguez-Jauregui E 1998, hep-ph/9804267; Mondragon A and Rodriguez-Jauregui E 1999, Phys. Rev. D **59**, 093009, hep- ph/9807214; Mondragon A and Rodriguez-Jauregui E 2000, Phys. Rev. D **61**, 113002, hep- ph/9906429; Kubo J 2004, Phys. Lett. B **578**, 156, hep-ph/0309167; Kubo J, Okada H and Sakamaki F 2004, Phys. Rev. D **70**, 036007 (2004), hep-ph/0402089; Beltran O F, Mondragon M and Rodriguez-Jauregui E 2009, J. Phys. Conf. Ser. 171, 012028; Morisi S and Peinado E 2010, 1001.2265.

[5] Kubo J, Mondragon A, Mondragon M and Rodriguez-Jauregui E 2003, Prog. Theor. Phys. **109**, 795, hep-ph/0302196; Felix O, Mondragon A, Mondragon M and Peinado E 2007, AIP Conf. Proc. 917, 383, hep-ph/0610061; Mondragon A, Mondragon M and Peinado 2007, Phys. Rev. D **76**, 076003, 0706.0354.

[6] Barger V, Logan H E and Shaughnessy G 2009, Phys. Rev. D **79**, 115018, 0902.0170. Gupta R S and Wells J D 2009, 0912.0267.

[7] Gunion J F, Haber H E, Kane G and Dawson S 1990, *The Higgs Hunter's Guide*, Addison-Wesley; Quigg C 1997, XXVII Int. Meeting on Fundamental Physics, Sierra Nevada (Granada), hep-ph/9905369; M. Spira and P.M. Zerwas, Int. Universitätswochen, Schladming, hep-ph/9803257; Ginzburg I F, Krawczyk M and Osland P 2002, hep-ph/0211371; Kanemura S, Okada Y, Senaha E and Yuan C P 2004, Phys. Rev. D **70**, 115002, hep-ph/0408364.

[8] Hoang L N 1998, Mod. Phys. Lett. A **13**, 1865, hep-ph/9711204; Adler S L 1999, Phys. Rev. D **60**, 015002, hep-ph/9901449; Ferreira P M and Silva J P 2008, Phys. Rev. D **78**, 116007, 0809.2788; Howl R and King S F 2009, 0908.2067.

[9] Barroso A, Ferreira P M and Santos R 2006, Afr. J. Math. Phys. **3**, 103, hep-ph/0507329; Barroso A, Ferreira P M, and Santos R 2006, PoS HEP2005, 337, hep-ph/0512037; Barroso A, Ferreira P M, Santos R and Silva J P 2006, Phys. Rev. D **74**, 085016, hep-ph/0608282.

[10] Konisi G and Saito T 1995, Prog. Theor. Phys. **93**, 1093; Toghi A and Moazzen M 2009, Int. J. Theor. Phys. 48, 3372.

AUTHOR INDEX

Institute of Physics
Dirac House, Temple Back
Bristol BS1 6BE UK

CURRAN ASSOCIATES INC
proceedings.com